Common Sense Transition

A Call to Action and
A Blueprint for Change

Karl Monger
Major (USAR – Retired)

www.KarlPMonger.com

Creative Team Publishing
Fort Worth, Texas

Disclaimers:
- Due diligence has been exercised to obtain written permission for use of references or quotes where required. Additional quotes or references are subject to Fair Use Doctrine. Where additional references or quotes require source credit, upon written certification that such claim is accurate, credit for use will be noted on this website: www.KarlPMonger.com.
- The opinions and conclusions expressed are solely of the author and/or the individuals represented, and are limited to the facts, experiences, and circumstances involved. Certain names and related circumstances have been changed to protect confidentiality. All stories where names are mentioned are used with the permission of the parties involved. Any resemblance to past or current people, places, circumstances, or events is purely coincidental.

Hardcover Edition

ISBN: 978-0-9967946-8-8

PUBLISHED BY CREATIVE TEAM PUBLISHING
www.CreativeTeamPublishing.com
Fort Worth, Texas
Printed in the United States of America

*****

Few are those who willingly go into harm's way
to keep us free.

Great are their personal and professional sacrifices.

They have earned our gratitude and respect.

We are committed to helping them live
with purpose and hope. Join us.

*****

For the purposes of this book, I use "soldier" or "veteran" in place of "service-member" or Marine, Sailor, Airman or Coastguardsman. I do this for flow and continuity and mean no disrespect or omission of recognition of the valuable service given my members of every branch of the service.

Dates of events are nearly as accurate as I recall, and might be off a week or two, and I have changed some names to protect the privacy of soldiers undergoing difficult times or involved in sensitive situations.

I also want to be upfront and say that although I have "credit" for a combat deployment during the closing phase of Desert Storm, I was never shot at, nor have I ever shot at anyone. I didn't do anything heroic, but I had the honor and privilege of serving beside and training heroes.

Endorsements on behalf of
Common Sense Transition

In the decade that I've watched Karl Monger mentor transitioning troops, I can tell you this: there is no one better equipped to help the motivated succeed. Karl puts all of himself into this book. If you are coming at your future with the intent of doing whatever it takes to get where you want to be, read this thing cover to cover.
  ~ **Nick Palmisciano, CEO Ranger Up**
  www.rangerup.com

Karl Monger has written a necessary and useful book, driven by his love and dedication to those who have served our nation. If I had the power to do so, I'd put a copy in the hands of every soldier as he or she leaves uniformed service.
  ~ **Lt. Gen. Jim Dubik, US Army, Retired, Author,** *Just War Reconsidered: Strategy, Ethics, and Theory*

Karl Monger <IS> the Ranger Shepherd. If it were not so, I am not here to write these words.
  ~ **US Army Ranger veteran and former CIA operative**

I've worked side-by-side with Karl Monger for nearly a decade. He's a mentor and a friend who *knows* what the hell he's talking about. I wholeheartedly endorse this book.
  ~ **Sgt. Boone Cutler, US Army Retired, Author,** *Voodoo in Sadr City*, **and Host of the nationally syndicated talk show, "The Tipping Point"**

Karl has written a book that asks the critical question of each and every one of us: if we loved our battle buddy enough to kill or die for them while we wore the uniform, is there any reason we can't love them enough to save their life after we take the uniform off? Sometimes it's as easy as introducing a veteran to someone and telling them, "Trust me, hire them" or maybe it is as difficult as driving hundreds of miles on a weekend to spend time with an emotionally wounded friend that is teetering on the edge. If leadership is about leading by example, I can tell you that my friend Karl is one of those rare guys that can say he has been there, done that both in uniform and out.

~ **Bill Cooper, former US Army Officer, GallantFew Founding Board Member, and COO of Bergkamp, Inc.**

Karl is one of the most revolutionary leaders I have met in the veteran space. His Blueprint for Change should be read by every leader in the military and veteran community.

~ **Mark J. Lucas, US Army Ranger veteran**

Karl Monger, founder of GallantFew, the outstanding revolutionary veteran network which has helped thousands of veterans transitioning to civilian life after serving our country, has created an essential tool for all veterans making the move from active duty to a successful life in the civilian sector with his new book, *Common Sense Transition: A Call to Action and a Blue Print for Change* proving that Rangers truly do Lead the Way! RLTW!

~ **Tim Abell, US Army Ranger veteran, actor, and GallantFew board member**

I can think of no more qualified person to write this much-needed book than Karl Monger. Karl initiated groundbreaking and effective veteran support for his Ranger brothers and beyond. His steadfast leadership in the Warfighter community has made a real impact.

~ **Michael Broderick, US Marine Corps veteran, actor and GallantFew board member**

Karl Monger has devoted every ounce of his being to assisting veterans with issues transitioning to civilian life and there is no one more knowledgeable than him on the causes and solutions to those problems. As a former airborne Ranger officer, Karl understands the mindset of the warrior and the realities of military life. As the founder and executive director of the veteran's organization GallantFew, he has helped veterans solve deeply personal and complicated problems by creating an innovative network of veterans nationwide to assist those veterans, and by teaming with varying specialists to fill the gaps that exist in the VA and other government veterans' programs.

The warrior mindset teaches soldiers to "suck it up and drive on." While there is strength to be gained from that rough and ready mindset, often times a veteran's decision to "suck it up and drive on" is a mistake, often masking a serious problem with drugs or alcohol.

As the Chairman of GallantFew's board of directors and a Ranger veteran myself, I have seen firsthand that Karl's thought leadership and execution have helped countless veterans with serious transitionary issues solve them in a positive environment and walk away equipped with the skills needed to flourish.

Every member of the armed services should read a copy of Karl's book when they exit, whether they use Karl's lessons to help themselves or to help their battle buddy.

~ **Rick Welsh, US Army Ranger veteran, attorney, and GallantFew Chairman of the Board**

Dedication

To those Gallant Few who have borne the burden

Of keeping this nation free

And who silently and without complaining

Carry the pain of physical, mental, and emotional injuries

Incurred while bearing that burden;

And to my sixteen friends

Who died in service to this nation.

Introduction

"Recognizing that I volunteered as a Ranger, fully knowing
the hazards..."
The Ranger Creed

The US Army Ranger is an elite warrior, driven to excellence and surrounded by a team of warriors striving to be better today than they were yesterday. The lessons shared in this book come from my experiences as a Ranger officer and from my frustrations, failures, and successes in transitioning to the civilian world.

Can you identify a positive term for transition from the military besides retirement? The military uses a sterile term, "Expiration Term of Service, or ETS". The most common terms are discharge and separation — implying something no longer needed or desired. This mindset sets the stage for some veterans who are never able to re-create the purpose they once felt serving their nation.

This book explores why that is and offers ways to help change this mindset. Most programs that help veterans are reactive in nature and don't step in to help until there is a crisis usually associated with the big three: unemployment, homelessness, and suicide.

We must move to a proactive, preventive approach and that must begin before ETS.

The individual replacement system didn't work well for soldiers going to combat in Vietnam. Give someone training in a sterile environment then send them off to "fly or die" on their own in an unknown land with unknown team-mates. There was a reason combat-hardened soldiers in Vietnam didn't want to make friends with the FNG (f***ing new guy), because they were going to make mistakes and get killed.

We're making the same mistake with soldiers returning to civilian life. We put them through a "transition assistance program" which is usually a two-week PowerPoint class of content approved by the government-specified contract which was awarded the lowest bidder several years ago. The content is the same whether the transitioning soldier is a private or a colonel and is taught by military retirees or spouses, who have little to no transition experience of their own.

Many of the transitioning soldiers still have daily responsibilities which prevent them from fully focusing on transition, not to mention the logistics of moving a household from one state to another. I've spoken to sergeants and captains who didn't take proper time to go through this training because their job responsibilities were "too important."

The result is that most plan on "figuring it out" when they get to wherever they are going to call home. They rapidly find out the situation on the ground is radically different than that portrayed in the transition training and they find themselves alone, isolated, with no buddies—and many of them make mistakes that at best increase the difficulty of a successful transition and at worst find them dead from the bottom of a bottle or from a self-imposed gunshot.

Exacerbating the situation is the fact that communities are not tied in to the process. Neither the military nor the Veterans Administration provides local communities with lists of soldiers returning home, leaving the soldiers to find their own way.

We fixed the individual replacement system by moving to unit-focused deployment, and we can fix the transition system by moving to a community-focused approach.

This book, therefore, is aimed at two primary audiences. The first: the soldier who is transitioning. My hope is that the lessons I've learned both through my own transition and from the experience of helping hundreds of others transition will resonate with the veterans going through that transition — whether that transition might start tomorrow or may have been ongoing for years.

As of the writing of this book, our nation has been at war for nearly sixteen years. Most of the men and women serving today volunteered after 9/11. They signed up knowing we are at war, and they signed up knowing veterans return from war and struggle with transition — yet they still do — "fully knowing the hazards."

The second audience is comprised of the communities that will become home for these warriors. You are likely a member of one of these communities, and with membership comes responsibility. Learning what you can do and doing it can make all the difference in a soldier's civilian life.

Communities deserve the benefit of the leadership and training our nation has invested in the men and women who served. Ensuring their transitions are successful will benefit the communities they call home.

Finally, I believe the principles and methods in this book can help anyone, not just military veterans. Imagine if you learned to live with intent—developing functional fitness in the Spiritual, Emotional, Physical, Social, and Professional areas of your life, and that you possessed the ability to choose your response to any given situation ("Response-Ability"). You may not be an Army Ranger, but these lessons learned will help you move towards success in any area of your life.

Foreword by Michael Schlitz
Sgt. 1st Class (Retired)

I first met Karl Monger in 2009 when a Ranger buddy I served with in Iraq told me about an older Ranger putting together a network of Rangers Nationwide. It wasn't long after that, in early 2010, that he was on a business trip in San Antonio with his business partner. I was still living there and going through treatment at Brooke Army Medical Center. We decided to meet in person for lunch and hit it off from the very beginning. He told me he was planning the Ranger reunion for the US Army Ranger Association to be held in September. I offered my assistance for any coordination that needed to be done in San Antonio area leading up to the reunion.

It was during that time I truly got to know Karl and his passion for giving back to the Ranger community and Veterans in general. There were many phone calls and emails throughout the year. He also told me about his idea for a non-profit called Gallantfew.org. The premise behind the nonprofit was similar to Big Brothers Big Sisters, pairing up the older generation Veterans with younger veterans to be able to network both professionally and socially. In 2011, we both hit the ground running to raise awareness about GallantFew and spent a lot of time traveling together.

I personally think I spent more time with Karl that year than his wife did.

I can say that I was not prepared to enter the business world after medically retiring from the Army. My transition was a little different because of the injuries I sustained in Iraq in February 2007. I lost both hands, suffered 85% burns to my body, leaving me disfigured, and I had severe vision loss. I physically had the ability to do the work, but I don't think I was mentally prepared enough for the differences in how we accomplish missions in the Military vs. how civilians do business. Karl Monger had an enormous impact on me and remains an inspiration through his leadership, mentorship, and guidance. Without him I don't think I would be as successful in my professional life and social life as I am today. The same tools, techniques, and practices described in this book are exactly what helped me.

I have witnessed Karl become a pillar in the Ranger and Veteran communities alike. He has completed several training courses and continues to seek education to be able to provide the best possible support for our transitioning veterans.

It is my hope that this book will be used by military leaders, soldiers, and veterans to better prepare for the rigors of transition back into the civilian world. This book provides first-hand accounts and personal stories of veterans who face the challenges of exiting the military and entering the civilian world. The tools, techniques, and practices used in this book have been proven by me and fellow veterans, and were keys in our own successful transitions.

I commend Karl Monger on his dedication in putting these common-sense practices together into a book that can be used by so many. I look forward to seeing the success of many transitioning veterans who will use this book as a tool as they enter new phases of their lives.

Karl Monger and Michael Schlitz, December 2011,
Texas Army National Guard Dining Out

Foreword by Kelly Burris, PhD, MBC

Functional Emotional Fitness™
Measurable Outcomes at Every Session
www.BurrisInstitute.com
www.BurrisConnect.com

Much More Than Military Transition

Common Sense Transition is not just a book for military vets and policy makers. It is for anyone who believes that we all must contribute to those who protect us. Most people do not realize there is a human wall between us and people whose only objective in life is to kill us. This is especially true of the human wall known as the Army Rangers and the special operations demographic Karl speaks of in his book. If you listen to the news as of late, special operations and special forces go in as the tip of the spear in our current conflicts. What Karl lays out in his book is the absolute least that should be done.

Karl's exemplary storytelling ability was immediately engaging. If you have ever wanted to get an inside view of the Army Rangers and special operations this is a good way to do it while learning how to contribute to these incredible entities. Admittedly some of the stories are difficult to read but I feel important for every civilian to know.

Three things stood out that Karl brought up about being a Ranger.
1. The Ranger Creed
2. The Four C's
3. Karl's Ranger Prayer

Karl's Ranger prayer exemplifies the commitment of this very special group.

> *Lord, let there be peace on earth – but if there is to be war, then send me first so others don't have to go.*

Another quote from the book that civilians like me have not considered: "A veteran is someone who has written a blank check to the country – up to and including his or her life."

Having been in the fight for veteran mental health since 2007, I am aware of the veteran plight regarding this issue. I did not know however the true extent of neglect which Karl clearly lays out. Issuing medication for instance that has known horrific side effects without any record of it being given to the soldier is one of those straightforward issues. No record means no recourse when physiological abnormalities arise later.

In the end, Karl lays out a very straightforward means of solving the complex problem of transition on a very grass roots level. It does, however, take the participation of the VA and the local community.

I can only hope that convoluted politics do not slow this process down. By being more informed through sources like the GallantFew.org we can make sure it does not.

Table of Contents

Table of Contents

Table of Contents

Chapter 1

A Tale of Two Rangers

Two Rangers both served in a Ranger battalion during different eras. Both received the same training, both went to combat, and both transitioned. There the similarities end.

The first parachuted into combat in Panama, during Operation Just Cause. Jumping at 500 feet into the dead blackness of night, "Mike" recalled seeing tracers rising from the ground and punching holes in the canopy of his parachute. He struck the ground with such force he "made a divot with his body." He broke bones in his leg and injured his back and shoulder in the impact, most likely blacking out for a few moments.

Regaining consciousness and fighting off the pain, he joined up with a buddy and fought through the night. The next day he saw a medic and told him, "Doc, I think I broke my leg on the jump." To which the Ranger medic replied, "Well, if you think you broke it, you're fine — move out."

Eventually the injury rendered him unable to fight and he was evacuated to the battalion aid station. Lying there with a broken leg, another soldier with fragmentary wounds was brought in and Mike held his hand while this soldier died.

How do you think he felt? He was ashamed and felt unworthy to be in the presence of this hero who had given his life when "all" Mike had was a broken leg.

Evacuated back, eventually returning to his home station, he was hailed at every stop as a hero, and with every accolade he felt more and more unworthy, the acute sense of failure clouding every thought. Now he was home for Christmas while the rest of his unit continued to conduct combat operations in Panama. The thought of his brothers continuing to fight and risk their lives while he was home safe and sound was unbearable. His self-imposed sense of failure was so acute that it resulted in his reassignment from the unit and ultimate discharge from the military.

He never asked the Veterans Administration (VA) to recognize the injury he received during the night parachute assault. For the next 25 years he fought physical and mental anguish, making a living serving as a military reservist and later as a handyman but never making more than just enough, and never feeling he had done his part, that he was good enough. Then a health scare and the accompanying financial burden brought him into our network. As I spoke with him on the phone my heart absolutely broke for this Ranger, who so deserved the benefits he earned when he jumped into combat.

When I learned where he lived, I realized there had been a group of Rangers who had met very near his home for a Saturday morning breakfast not two weeks earlier. I asked if he had known about the event and he replied he did, but that he didn't go, saying, "I didn't think I'd be welcome."

The reality is a young Ranger with combat tours in Iraq and Afghanistan holds the Ranger that jumped into Panama with the same esteem that Ranger holds for those who climbed Pointe du Hoc in World War II.

Combat soldiers, especially Rangers, are hard individuals. We impose stricter standards on ourselves than anyone else would. Would Mike's life have been different if another Ranger had sat down with him and helped him process this experience and accept the fact that his injury was as legitimate as if he'd been shot? Would it have been different if he'd received a disability rating from the VA that would have provided him the ability to pay basic bills rather than scrambling for just enough to get by?

I believe Mike's life was a shadow of what it could have been. It didn't have to be that way.

*****

In 2013 I received a message through LinkedIn from a young Ranger named Steven Barber. Steven had multiple combat tours and was leaving the Army to return home to Texas. He had heard about my Ranger network cultivated through GallantFew and eagerly sought assistance.

Steven and I exchanged emails and had a phone call in advance of his arrival back in Texas, and when he got here I took him to lunch. As we talked about his plan, he told me he had thought about being an insurance agent. Now, going from combat Ranger to insurance agent isn't a common path. I suggested there might be some alternatives. So I took him to my Rotary Club breakfast and several Chamber of

Commerce social events. I introduced him to as many local business people as I could. I also shared with him many of my own personal lessons learned over twenty years of transitioning.

Over the next two years, we met occasionally for lunch or he visited the Rotary club, and he kept me updated on his progress working towards his insurance certification. He was focused, and he was busy with a growing family.

Then in October of 2016 Steven called me up, and he proudly announced that the following Saturday would be the grand opening of the Steven Barber State Farm insurance agency! As I congratulated him on this achievement, he told me he wanted me to know how instrumental I had been in this. Stunned, I asked how that could be possible, and told me that be welcoming him, sharing my transition lessons, and introducing him to people in the community had given him a local foundation, a network of contacts that gave him the confidence to take a chance on his insurance dream, because he knew he had options.

Imagine that—such a simple thing. Connecting, introducing, including. Now the Steven Barber State Farm agency employs four people. A great success story for transition and indicative of what one can achieve with energy, motivation, a plan, and local support.

How different might Mike's life had been if twenty years ago someone with his background had welcomed him, got to know him and shared the lessons of transition so hard learned? I suspect the story would have been more like Steven's.

The last time I checked with Mike, the VA still had not recognized and approved disability for the injuries he had sustained parachuting at night, under fire, into a foreign land.

This led me to pen the following thoughts on a Facebook page for the 75th Ranger Regiment veterans. The comments that came back ran the gamut from, "Wow, this describes me" to "Rangers should be proud of who they are and not let bad things affect them" and even "I've avoided those I served with because of the way I was treated."

My post:

> *I have a theory, interested in your thoughts. There are basically three honorable ways to leave Regiment: Retired because you're freaking old; promoted or selected for a unit you compete for; or catastrophic injury (amputation, vision, etc.). Reasons like getting out to save a marriage, go to school, or any of the myriad other reasons (DUI – Driving Under the Influence being one) cause the departing Ranger to carry a sense of loss, failure, even shame. Especially if it's a wound or injury that's not externally visible.*

> *Consider the young Ranger that enlists on a Ranger contract, excels at basic, Airborne, and RASP (Ranger Assessment and Selection Program) and goes on to multiple combat deployments. He is pegged as a leader and sent to Ranger School – where he fails to get the tab, due to any number of reasons. When he returns, he's gone. Sent to the 82nd or other unit, or he ETSs (Expiration Term of*

Service). That Ranger now looks back on his Ranger time with embarrassment rather than pride and avoids the Ranger veteran community because he doesn't want to confront that shame, that failure.

It's compounded if a Ranger buddy sub-sequently dies or is seriously wounded; now add survivor guilt to the mix. I believe the Rangers that need this community the most are the ones that avoid it the most. We need to make a conscious effort to reach out to the guys we know that are not in contact, not on social media – they may build the hell in their minds to the point where it ruins the rest of their lives, and a simple "brother, how's it going" just might be a life-saving beacon.

I'll come back to this later—we know Rangers as well as other soldiers are going to experience these things, but we fail to warn them—to share our hard-won lessons learned, so they can accelerate into their civilian lives rather than drag the weight of failure into every waking thought.

I also have to insert here that while I was writing this book, Mike called me up. He told me that he had just got off the phone with the VA and that the VA had awarded him a 100% disability rating and were back paying him the nearly eight months to the date he had applied (they don't compensate from date of injury, only from when the veteran actually turned in the request). With emotion in his voice, he said "I just thank God for you ... you gave me some light at the end of the tunnel ... just knowing you guys were there for us ...gave me the motivation to get through the process, and knowing you had my back." He went on to say that as soon as he got his check, he would repay the grant I had sent him, and that he would double the amount because he wanted me to help another.

Then while in final editing of this book, I learned that an officer I had served with, who I ate breakfast with a mere eighteen months ago, ended his life. He was over fifty, was funny, handsome, a great guy. He suffered some setbacks in his career and in his relationships, and he never reached out.

Chapter 2

A Common Ranger

A young infantryman earned his Combat Infantryman Badge and a Presidential Unit Citation in the Philippines during the close of World War II. He went on to serve in three Korean War campaigns and ended his twenty-year career in Germany, retiring from the Army as a sergeant first class.

Thirteen years later, in 1977, he died alone, an alcoholic in a men's shelter in Denver, Colorado. When his son went to get his possessions, they fit in a shoe box. The son wasn't with him when he died. Their relationship was rocky, the son thrown out of the house when he was sixteen years old.

He, too, joined the military, ultimately retiring as a sergeant major. His son in turn, never knew him either, as divorce split that family apart when the boy was four. He had a two-year-old sister and his mom was pregnant. The impact of what was most likely untreated post-traumatic stress, probable traumatic brain injury, and alcoholism was now a terrible expanding funnel tearing through future generations.

Five years later a police officer told the boy's mother about a new program called Big Brothers, formed to help offset the growing trend of single mothers raising boys. He told her that young men without fathers were more likely to get in trouble and that this program appeared to help them. He became one of the first "Little Brothers" in America. He went on to mentor an at-risk youth when he went to college, and after serving in the Army, returned to run the agency that had been formative in his life so many years earlier.

This book on transition incorporates a lot of lessons he learned along the way. He is by no means perfect, having also gone through a divorce that devastated two teenage daughters. This book is very personal and much of it is his transition story and the lessons learned from helping others through their transitions.

This is my story.

I'm telling it to illustrate the character qualities that are developed in military service—the leadership skills, the amazing opportunities for responsibility, the ultimate risk of serving, and the value a veteran can bring to a community by seeking purpose and leadership roles. Everything that has happened to me seems to have a divine guiding purpose.

I had to grow up in a hurry. At five years old, I was becoming the man of the house. We packed up and moved from Manhattan, Kansas where my father was stationed at Fort Riley, and went to live with my mother's parents, Karl and Juanita Drowatzky in Pennsylvania. Juanita's brother, Roy Jones, served as a corporal and infantry squad leader in the 89th Division in World War I and years later our infantry experiences would create a bond that no one else in the family shared.

My grandfather, Karl, stood in as a father figure for me, and we had a life-long close relationship. He was the best man I have ever known. His brother, Frank Drowatzky, had served as a Cavalry scout in Africa in World War II, and as a kid, I never understood why he lived in his sister's basement. He was the grumpiest person I ever knew.

Years later in family documents I found a biography he had written and it was clear he was intensely proud of his service. The documents included a copy of the telegram sent **to** his parents (my great-grandparents) informing them that their son had been wounded in battle. I also found a history of his struggle with the VA, a log of the pain he experienced (he carried a lot of shrapnel in his body), most likely combined with undiagnosed post-traumatic stress and traumatic brain injury (this will be a pattern I see repeatedly in combat veterans). He died before I entered the Army, a bitter, broken, forgotten old soldier. A hero who lived in his sister's basement.

Mom and my two sisters, Kimberly and Kristin, lived with my grandparents for several years, moving to Sioux City, Iowa and then to Wichita, Kansas as my grandpa changed jobs. Mom became a teacher, working during the day and taking classes at night to get her degree. Many nights my sisters and I sat in the corridor at Wichita State University drawing pictures and trying to stay occupied while mom attended class. As soon as she could afford it, we moved from my grandparents' home into a home of our own.

I remember telling my elementary school classmates that my father was a test pilot and had been killed in a crash. I didn't know how to tell them he had left and I hadn't seen him since that day. I'm not going to place blame on anyone for why he wasn't there — I'll only remark that he wasn't. That's a hard thing to understand when you're six years old and everyone else has a mom and a dad.

By the time I was near ten, I was preparing food for mom and my sisters; she told me years later about finding her lunch in the fridge with a note from me. I don't remember that, but I do remember a gallon of milk falling and spilling all over the floor, and my mom sinking to her knees in tears, because there would be no more milk until she got paid again. It's difficult now to remember the despair. What I do remember is the love she showed and the fun we had as a family. I can't imagine how she felt, taking on all the responsibility herself, while at the same time continuing her education.

In the late '60s as divorce became more common and larger numbers of single parent mothers struggled with raising boys, the Big Brothers program came to Wichita. I became one of the first ten kids matched in the program, and Wichita attorney named Dave became my big brother. Every Tuesday, Dave came to pick me up in his white convertible Corvette. We went to movies, built and flew model planes, and played catch. He didn't have to, but he cared about me. Years later I learned that even my sisters were impacted by the fact that someone outside our family cared and spent time. Dave and I were matched for nearly three years, until my mom remarried.

My step-father soon became my legal father. In 1972 Richard Monger adopted Kimberly, Kristin, and I and we all became Mongers. Richard was a World War II veteran, enlisting in the Army Air Corps out of Chicago. He was a Private First Class and in training in San Antonio when the war ended and he was subsequently discharged. Richard was a good man, active in his church and solid in his faith. He and my mom were married for over forty years, until he passed away from congestive heart failure.

Typical of many in his generation, he never sought anything from the VA. When he was 80, at my urging he went to the Wichita VA hospital to enroll for prescription benefits—I walked in with him holding his discharge papers from 1945. We were turned away, no benefits. If he had signed up before the mid-80s, the VA clerk explained, he would have received benefits right away, but the laws changed then and there were no longer automatic benefits for World War II veterans. I encouraged him to appeal, but he said no; he'd tried and really didn't feel comfortable trying— as he "really didn't do anything in the war."

Richard F. Monger, US Army Air Corps, 1945

By the time mom remarried, I was at my fourth elementary school. Thankfully, things stabilized, and we lived in the same house from my sixth-grade year until I graduated from college and left home. I held my first job when I was in the seventh grade, as a paperboy for the Wichita Eagle and Beacon. It was in junior high school that I met my best friend, Bill Cooper, who would be so important in my transition later in life. I also was physically bullied by a kid who loved to fight, which was complicated by the fact that my mom led a very religious life, one which demanded that I "turn the other cheek." I had to walk to school right past the bully's house, so I got creative about walking routes to avoid the conflict.

In high school, I sang in a madrigal group, ran track, and made the gymnastics team. My mom absolutely forbid me to play football—which was terribly embarrassing to me then, but in retrospect the physical skills I gained in gymnastics helped me with obstacle courses and rope climbing in the Army, and I never had the knee problems so many of my football player friends had. I always regretted, however, not being part of the tight knit team that was the Southeast High School football team.

Perhaps a result of being bullied, I teased a kid in our class that was super smart but less gifted athletically. It's something I've regretted ever since, and at my high school twentieth reunion I sought him out and with tears in my eyes I asked his forgiveness. We've been friends ever since.

I went on to college less as a plan, but more because I didn't know what else to do, and it seemed like all my friends were automatically college-bound. I pledged a fraternity, and became a member of Beta Theta Pi. This was a group of young men that captured most of the athletic and scholastic awards on campus, and I soon found myself enjoying girls and beer more than I did my studies.

As my first semester ended, I saw a sign on the wall of the Engineering school. It advertised a "Marksmanship" class. I needed an easy A, and saw this as a way to bump up my GPA and have some fun at the same time. Little did I know that this was an ROTC class.

Soon I found myself in an underground firing range, beneath an old auditorium at Wichita State University (WSU). I doubt very many people outside the ROTC folks even knew of its existence, but there I learned the basics of marksmanship with a match grade .22 rifle under the watchful eye of Sgt. Maj. Seals.

Seals was a crusty old tanker, a Vietnam veteran that fulfilled virtually every stereotype of a senior non-commissioned officer. He was gruff, took crap from no one, and made sure that he was always teaching. He was a perfect fit to the Professor of Military Science, Lt. Col. Andrew Kushner. Kushner was a commanding figure, tall, square-jawed and had served in combat with the 82nd Airborne and a special forces unit. Sadly, he would pass away from cancer after he had retired and moved to his dream home in Alaska. He wore on his shoulder a little black and gold tab with the word "Ranger." He fostered a small group of cadets into a group called the "Wichita Rangers" and we studied our Ranger Handbooks in more detail than we did our textbooks. On weekends, we practiced patrolling and small unit tactics along the Arkansas River, and learned rappelling and rock climbing.

We obtained permission to rappel from the back of the university stadium bleachers. Going over the top of the bleachers meant facing a hundred-foot drop to the parking lot with the last two-thirds a free-rappel, meaning no wall to walk down. I learned how to do this "Australian" style, meaning face first, running down the wall with the rope streaming behind.

Part way through the semester, Kushner pulled me aside and suggested I consider applying for an ROTC scholarship. I did, and the Army paid for the next three years of college. My buddy, Bill Cooper, came back to Wichita from out of town, and also entered the ROTC program. Our friendship further cemented. During this time, I also volunteered as a mentor in what had become Big Brothers Big Sisters and was matched to an at-risk youth for several years.

At the end of my sophomore year, one of the cadre approached and offered me a slot to the US Army Airborne School. I now qualified for an official Army course as I had signed my scholarship contract, in effect an enlistment. I had not, however, attended any formal military training. Airborne School would be my basic training. Held at Fort Benning, Georgia, Airborne School was a three-week, physically demanding course, and in August 1981, the highs were in the 90s and the humidity averaged over 80%. The Army would not switch to running shoes and exercise clothing for several more years, so all physical training was conducted in combat boots and fatigues.

I had no idea what I had gotten myself into. As I walked up the stairs to the barracks at the 42nd Company, 4th Student Battalion TSB (ABN), I was stopped by a captain, shoulders wide as the door I was about to enter. "Cadet!" he screamed, "Get a &$%#! haircut. RIGHT NOW!"

Heck, I was a college student. Probably the longest hair of my life, over my ears and touching my collar. Thanks ROTC cadre, for making sure I was prepared! After a few more choice words from the captain, I was on my way to get my first military haircut. Didn't matter what you told the barber, everyone in that seat got the same cut. Razor set to the lowest possible setting, head shaved. It is a bizarre feeling, having no hair.

Returning to the company area and feeling a bit more anonymous, I found my assigned room. The August Airborne classes are prime time for cadets to attend, and this cycle was no exception. My roommate was an Annapolis cadet. I said hello and stuffed my bag into my wall locker, which was more of a cabinet than a locker, with space on one side to hang clothing and storage drawers on the other side. I slammed the door shut and turned around to see my roommate's expression of horror. "What are you doing?" he demanded. "You have to hang up your clothes!"

After another trip to the shoppette, this time to buy hangers (not provided!), I returned to my room. A thoroughly disgusted midshipman spent the next hour teaching me how to properly hang my gear, make a rack, and spit-shine my boots. I like to think I caught on quickly, but I'm sure it was the last thing he wanted to do. I still remember my phone call to my grandparents that evening.

After standing in line for a payphone, I had five minutes for a call. It's amazing how much you want to talk with someone on the phone after you've left your safe comfortable familiar home for the first time. My grandmother was also astounded that the Army didn't provide hangers.

The next morning, at zero or "0-dark thirty," I learned the command "Half-left, face!" and subsequent "Front leaning rest position, move!" It seemed nothing we did, either collectively or individually, was correct. We did pushups for everything. First the administrative cadre put us in formation and accounted for us; then they turned us over to the "Black Hats," the famed Airborne school cadre, by marching us from the company area to the ground week training area. This was my first experience with "Jodie call," or a cadence sung in time to marching. It was thrilling, hearing the deep voices of a hundred soldiers echoing off the barracks in the pre-dawn darkness.

The reception area was a gravel lot containing steel cables arranged in straight lines, very difficult on which to march. Our cadre marched us onto the lot so we lined up on the cables then turned us over to the Black Hats. Distinctive for their sharp, black ball caps with shiny golden rank and silver wings gleaming, the Black Hats have been responsible for turning mobs of soldiers and future soldiers into Airborne troops for over fifty years.

The Black Hats began our day by conducting inspections. An inspection consists of a sergeant standing nose to nose with you while he picks apart any stray thread, wrinkle, piece of lint, or scuffed boot. Any infraction meant being sent to the "gig" pit where another sergeant waited to put you through the 80s version of cross-fit. I believe I reported to the gig pit every day while I was there, once for not shaving the back of my neck.

I also learned about sick call. Not that I went, but I was astounded how many soldiers reported to sick call the second morning of training. After coming all this way, and being given the chance to earn my wings, I wasn't going to let anything prevent me from completing the course.

But enough about jump school. Three weeks and five jumps later I felt like the proverbial round peg in a round hole. As I stood at graduation and asked the sergeant to give me the privilege of blood wings I felt more proud that any other time in my life. Being sent to Airborne School as an inexperienced, untrained cadet was a risk for my school's cadre and would not have reflected well on them had I failed to graduate—but the result was a motivated cadet who returned to college intent on becoming an infantry officer and attending Ranger School.

The summer before my senior year in college I attended ROTC Advanced Camp, in effect basic training for ROTC cadets. At one of the ranges a sergeant called me out of the bleachers and told me a sergeant major wanted to see me. I walked around the back of the bleachers and facing me was my biological father. It was surreal. This was the first time I'd had contact with him since I was four, and here in front of me was an exact carbon copy of me. It was like Marty McFly in the 1985 movie *Back to the Future* meeting his ancestor. Except for the snow-white hair and a few more pounds, it was me.

He took me out to dinner, and it was an emotional evening for me. Since then I've seen him perhaps a dozen times, and he has a full life with three other kids and grandkids. Long ago, Richard Monger became my father, but this man is more like a distant cousin. How much of all that happened, of that which tore my family apart, is attributable to the failed transition of my biological grandfather? The question is unanswerable — but it provides me motivation to prevent the possibility of it happening to the child of another soldier by helping him or her move from military service to a civilian life of purpose and hope.

I graduated from WSU in May 1983, was named senior honor man from Beta Theta Pi, with a degree in Administration of Justice and a minor in Political Science. Anyone that asked me my degree got the same answer: "Second Lieutenant." The very same day as graduation, I received a Regular Army commission as a 2nd Lt. in the Infantry — a Regular Army commission being an honor reserved for graduates of West Point and the Distinguished Military Graduate from each ROTC program.

Bill was commissioned the same day alongside me, but he had lost some credits when he transferred to WSU, so he didn't go on active duty until a year later. He did agree to be my co-pilot as I drove cross-country (at a max speed of 55 mph thanks to President Carter) to report in to Fort Benning. The lump in my throat as I drove away from home developed into a soreness that kept me hoarse for a month.

"Lead, Follow, or Get the Hell Out of the Way" read the sign over the admin offices for the Infantry Officer Basic Course (IOBC). A sixteen-week course, IOBC was the rite of passage to become an Infantry Officer and wear the blue shoulder cord of the Infantry. It was there I learned land navigation, road marching, basic tactics, and saw true character under hardship and got my first glimpse of self-serving and cowardly behavior.

About half-way through the course, we had a presentation about Ranger School and the instructor passed out forms which we had to fill out and sign to volunteer for Ranger School. Despite a full colonel standing in front of a class of over 200 young new infantry officers telling us that a real leader wears a Ranger tab, less than half signed the form volunteering to go.

Chapter 3

The Army's Toughest Leadership School

Eight long, sleepless, and hungry weeks comprised Ranger School. Officially the Army's "toughest leadership school," Ranger School intentionally deprives students of sleep and food to simulate the stress of combat. Students are then rated against and by each other. My class started with over 200 prospective Rangers and less than half of us completed the course without recycling. Among them was a young 2nd Lt. Stefan Banach with whom I served as a captain at the 1st Ranger Battalion and who was 3rd Ranger Battalion Commander on 9/11. He earned his place in history in October of 2001 when the invasion of Afghanistan began. I also became friends with Paul Long, who became the first of fifteen of my friends to die on active duty. He was killed just before Christmas 1985 in the explosion of an aircraft bringing a peacekeeping force from the 101st Airborne Division home from the Sinai.

Two buddies in Ranger School who ate together, trained together, graduated together — one will forever be remembered in Ranger lore and history for his part in the Global War on Terror, the other, who died too young, sacrificed for our freedom before his full potential was realized.

Ranger School revolves around the buddy system. If you are not with your Ranger buddy, you are flat wrong. Your Ranger buddy watches your back and you watch his. My Ranger buddy was an Air Force Combat Controller (rare for an airman to attend Ranger School) named Charles Granda. Charlie was strong as an ox, with an easy grin and outgoing personality, and I owe my Ranger tab to him.

One of the requirements to earn that black and gold piece of cloth is successful completion of the Darby Queen, twenty-six obstacles spread over a two-mile long world class obstacle course that meanders up and down hills in a forest on Fort Benning. Students are given three chances to complete an obstacle. Failure to complete two obstacles equals recycle to the next class or drop from the course.

Remember my gymnastic background? Obstacle courses were more fun to me than a challenge, and when I got to the overhead ladder I tackled it with ease. The overhead ladder consists of wooden rungs that are about six inches in diameter—too large to wrap your hand around, you must keep an open grip and keep moving. It's still tough, but I breezed through it, tapping the last rung and dropping to the ground. "You didn't complete that obstacle, Ranger," a Ranger Instructor (RI) called out, so I ran back to the start of the ladder and began again. I made it halfway before my arms gave out and I fell off. I shook out my arms and returned to the beginning, and this time Charlie grabbed my legs and tried to help me along—but I was too burned out. Again, I dropped off; the RI recorded my roster number, and directed us to continue.

The very next obstacle required us to climb a rope about 40′ tall to a log walk then slide down to the ground. Charlie went up first, having no trouble. He stood at the top of the rope, both feet planted firmly on the log, and yelled encouragement down to me. I climbed over halfway up the rope and it was all I could do to hang on. Ascending higher was not going to happen. Charlie yelled at me to hold on and he reached down and began hauling the rope, and me, up — hand over hand. When I got to the top, he grabbed my wrist and pulled me atop the log. We ran across the top and slid to the ground. I looked at the RI and he pointed to the next obstacle. We completed the remainder of the Darby Queen with no problems, crossing the finish line stride for stride.

Charlie remains a good friend of mine today — after the military, we fell out of contact for years but thankfully we reconnected around 2010. I owe him a tremendous debt, not the least of which is my tab. By the way, he was also a genius at hiding small pieces of candy (prohibited by the RIs) and giving me one from time to time.

Ranger School places tremendous physical demands on a person. The Army still conducted all physical training in boots and fatigues, and Ranger School was no exception. The required five-mile runs in formation and the two-mile runs for time, all done in boots. The wear and tear on feet, knees, hips, and backs is staggering and we often joke that Ranger years are like dog years — one equals seven.

One of my proudest memories of Ranger School is from rappelling in the mountains in northern Georgia. As an ROTC cadet I had become an instructor in rappelling and we often free-rappelled from the back of the school football stadium, over a hundred feet of open air underneath the stadium overhang. I loved the thrill of screaming down the rope, and this day was no different. I bounded off the top of the rock ledge and let the rope slip through my open hand, locking in the brake about twenty feet above the ground, just enough that the stretch of the rope would allow my feet to touch the ground. Releasing the rope at exactly the right time, it looked like one continuous jump to an easy landing. An RI walked up to me and asked which battalion I was from, and I saw the disappointment on his face when I told him I was a brand spanking new 2nd Lt.

I got an honor grad "go" (it is a go/no-go graded system) on the first patrol I led, and I got a major minus spot report (two will get you recycled) on my next when my entire ambush fell asleep. I got angry at a Ranger who fell asleep in front of me during a movement in the mountains and kicked him as hard as I could—only to discover how hard a tree stump can be.

In Mountain Phase, we awaited helicopters to come pick us up for a lengthy stay in the mountains of northern Georgia. As we sat waiting in a "stick"—a line of soldiers designated for a particular aircraft—the Ranger Instructor casually asked us if we'd like to know what was going on in the world. "Sure," we replied, because a Ranger student is isolated from most outside news. The day prior, he told us, the 1st and 2nd Ranger Battalions had parachuted onto a runway on the island of Grenada, freeing American students. "Which of you are second lieutenants?" he asked. I and a few others raised our hands. "You better pay attention," he advised, "because Ranger lieutenants are dying in Grenada and they're going to need more." While no Ranger officers died in combat in Grenada, it was a sobering realization.

A few weeks later, we moved on to Desert Phase. My class was the second to go through this new phase in Ranger School, adding to the traditional Benning, Mountain, and Jungle Phases.

We flew from Georgia to Fort Bliss, Texas and parachuted onto a dry lake bed. We were so tired we kept falling asleep during the jump commands as we neared just moments from exiting the aircraft. At the last minute, one of the Ranger Instructors moved me from one side of the aircraft to the other, and when I exited the bird, my right arm was nearly yanked out of its socket. The static line that deploys the parachute had wrapped around my arm. This is not an uncommon jump injury and the usual result is a condition called "Popeye arm," where the bicep is ripped from the upper arm and bunches up around the elbow. I was very fortunate in that I was wearing a field jacket liner under my jungle fatigue shirt, a lightly insulated top made of quilted nylon. The slick material allowed the static line to slip around the bicep and I ended up with a torn muscle that still today looks like someone took a chunk out with a spoon. I was in tremendous pain but was determined not to go on sick call and risk being recycled or dropped from the course. I persuaded the medics to give me some analgesic horse pills to help me tolerate the pain. Luckily, we were past the physical harassment portion of the course; if I'd been dropped for pushups, I would have been lucky to do one.

Because I never sought official care for this, there is no record of the injury in my file — even though I recall being interviewed by a safety officer who concluded the injury was a result of "jumper error." The VA denied my request for recognition of this service-connected injury.

Every Ranger student is required to memorize and recite on demand the Ranger Creed. The Creed has been part of my life ever since:

> *Recognizing that I volunteered as a Ranger, fully knowing the hazards of my chosen profession, I will always endeavor to uphold the prestige, honor and high esprit-de-corps of my Ranger Regiment.*
>
> *Acknowledging the fact that a Ranger is a more elite soldier, who arrives at the cutting edge of battle by land, sea, or air, I accept the fact that as a Ranger my country expects me to move further, faster and fight harder than any other soldier.*
>
> *Never shall I fail my comrades. I will always keep myself physically fit, morally straight and shoulder more than my share of the task whatever it may be. One hundred percent and then some.*
>
> *Gallantly will I show the world that I am a specially selected and well-trained soldier. My courtesy to superior officers, neatness of dress and care of equipment shall set the example for others to follow.*
>
> *Energetically will I meet the enemies of my country. I shall defeat them on the field of battle for*

I am better trained and will fight with all my might. Surrender is not a Ranger word. I will never fall into the hands of the enemy and under no circumstances will I ever embarrass my country.

Readily will I display the intestinal fortitude required to fight on to the Ranger objective and complete the mission, though I be the lone survivor.

Rangers lead the way!

The Ranger School overall graduate rate is forty percent, and that includes students who recycle. Less than thirty percent go straight through, and I'm very proud that I was one of those who did.

My parents made the drive from Kansas to Georgia for Ranger graduation, and my mom pinned the black and gold piece of cloth on my uniform. She brought with her boxes and boxes of cookies and other goodies, and on our way out of town I made them stop at McDonalds, Kentucky Fried Chicken, and Pizza Hut. I ate until I passed out.

My first operational assignment was as a platoon leader in the 2nd Battalion, 60th Infantry, 9th Infantry Division. 2/60 had earned its motto "Scouts Out" in combat in WWII. My battalion commander was Lt. Col. Glynn Hale, a future member of the Ranger Hall of Fame. My company commander was Capt. Randy Bell. We were part of the 3rd Brigade, which was designated as the Army's "High Tech Test Bed." That meant our scouts ran around in cool dune buggies, and although my platoon consisted of M901 ITVs (the M113 Armored Personnel Carrier modified with a hydraulic anti-tank missile system), we pretended they were

LAV-25s, six-wheeled vehicles with 25 mm cannons. It made planning defenses and creating range cards challenging as our mission was to test out the concept of what was to become the Stryker concept, ubiquitous in combat decades later in Iraq. During this time, we also tested out the first GPS systems, successfully marking imaginary minefields during a breaching operation at the National Training Center.

Shortly after reporting to the unit, we went to the field on tactical exercises that involved staying in the woods for several days. When the supply sergeant set up the chow line — thermal insulated containers containing mashed potatoes, sliced ham, green beans, and a large container of iced tea — the soldiers lined up to get their food. As I walked up, my company commander Capt. Randy Bell motioned me over. He said simply, "Leaders eat last."

The captain went on to explain that out of a show of respect for the soldiers who carry the heaviest loads, who put themselves in the greatest harm's way, leaders eat last. He pointed out to me how the corporals in charge of four soldiers let their four go first. The sergeants in charge of the corporals waited until their two corporals and their soldiers went through the line, then the sergeants went through. This went on through the ranks until all the enlisted soldiers received their food, then the captain motioned for me to go before him.

This tradition accomplished several things. It ensured the quality of food remained high until all hundred plus soldiers had eaten, and the captain at the end knew that everyone before him had received enough, and that the quality was at least as good as he received, if not better. It demonstrated to the lowest ranking soldiers that their leaders care about them.

Leaders in the army have many responsibilities and demands on their time. The decisions they make determine who lives and who dies. They certainly could pull rank, eat first and get on to their important tasks—but good leaders realize the most important thing they can do is to earn the respect of those they lead by putting those they lead first. Imagine our society if that was the norm.

My stint as a rifle platoon leader lasted a short four months then I was "promoted" to be the support platoon leader. My battalion commander wrote on my evaluation "2LT Monger is the best platoon leader in the battalion." I put quotations on "promoted" because the support platoon I took over was a flat mess. A mechanized unit burns through fuel, food, and ammunition, and it's the support platoon that delivers the means for the battalion to carry on the fight. It was a huge increase in responsibility, and one with a steep learning curve. When I assumed responsibilities for the support platoon, only four of the twenty supply trucks were road-worthy and it was common knowledge that morale in the small unit was low.

At the time, I would much rather have had the sexy job of scout platoon leader, and run around the battlefield in high tech dune buggies. In retrospect, I gained a tremendous amount of knowledge and the experience of putting this team back together provided me skills I've used repeatedly since.

Shortly after I took over the support platoon, I put in place an aggressive plan to increase the maintenance level of the fleet combined with tactical field training — how to drive a five-ton truck at night with no headlights (and without night vision gear) while maintaining proper spacing in a convoy.

Late at night we were crossing a local highway that cut through the Fort Lewis training area. I had my driver park my M-151 jeep on the side of the hardball with headlights on while the massive trucks crossed one by one. I placed myself in the center of the asphalt so I could see both ways down the road and I waved each truck across. The fourth truck in the convoy was pulling a cargo trailer. As it crossed the road, through the gap between truck bed and trailer I saw headlights far up the road. In a flash there was an explosion, the trailer flew up and to my right, spinning past me. A chrome trim strip from the car hit my leg as a Camaro flew within a foot of me, careening down the road and stopping in a ditch, the roof peeled back like a tin can. I ran to the wreckage, shouting for all vehicles to turn on their headlights. When I reached the driver's door I saw the man's head was a bloody mess. I felt for and confirmed a pulse and a sergeant run up to me. I told him to stay with the victim and keep him breathing.

I ran back to my jeep to call for help and as I did, a patrol car pulled up, the officer inside asking if we had seen a Camaro go through as one had just run through the post checkpoint without stopping.

I learned later that the soldier driving the Camaro survived but required extensive dental reconstructive surgery. His blood alcohol content was more than double the legal limit. My brigade commander ordered an Article 15-6 investigation, and the captain that conducted the investigation found me without fault. My brigade commander disagreed, and issued me a letter of reprimand and fined me a month's pay.

This could have derailed my budding military career, but I was determined to repair my reputation, and with the support of my battalion chain of command, I was given the leeway to plan and conduct training as I saw fit.

Soon the platoon got a new platoon sergeant. Sgt. 1st Class Julius Kimmie was a consummate professional and together we rebuilt put the platoon back into an effective team. Within the year we won the title of Best Support Platoon in the Brigade during a surprise roll-out inspection of driving skills, mechanical performance, and documentation. When I left the platoon a year later to become a company Executive Officer, I was very proud of what we had accomplished together. I was also the proud recipient of the Expert Infantryman's Badge, the only officer in the Brigade to earn it that year along with less than a dozen enlisted soldiers.

Ironically, the brigade commander that had issued me that letter of reprimand was the same officer to pin the blue metal badge with a silver rifle on my fatigue uniform, still damp with sweat from the timed twelve-mile march with gear and rifle, the final trial of a week-long test of combat skills. Col. Barry McCaffrey went on to gain fame during Desert Storm by driving the 24th Infantry Division to success, subsequently serving as President Clinton's Drug Czar.

Chapter 4

Stark Reality

In 1985, the stark reality of the seriousness of my profession hit me. Arrow Air Flight 1285 carrying 248 soldiers home from peace-keeping duties in the Sinai, exploded over Gander Newfoundland. Never proven, many believe the bird was sabotaged with a bomb timed to explode in flight. As I looked over the casualty list, I recognized the name of my friend 1st Lt. Paul Long. Since new Lts at the Infantry School lined up alphabetically, Long stood next to Monger. We became friends and went to Ranger School together. Paul was from Arkansas, was a quiet, solid professional officer. He was first of fifteen of my friends to be killed on active duty.

Happily, I also became a father. I had taken leave in the spring of 1984, returned home to Wichita and married Jana Campbell. Jana and I had dated my last two years of college and in 1985 our daughter, Jackie was born. Jana was a wonderful mom, and she did the heavy lifting of parenting as the Army demanded more than all my free time.

After three years at Fort Lewis, it was time to move. We learned of our future through an official envelope delivered to our quarters, ordering me back to Fort Benning to attend the Infantry Officers Advanced Course, after which we were to go to Hawaii. I called my mom to tell her news, and told her we'd been assigned to paradise. She wanted to know if that was Fort Riley! Now as a grandparent I know how difficult it is to be so far apart — but at the time, Fort Riley was anything but paradise to us.

At the Advanced Course, we learned the details of the military tactical decision-making cycle, how to take an assigned mission and unpack the specified and implied tasks, develop and wargame courses of action, and select and implement them. Upon graduation, I attended Air Ground Operations School, a short three weeks at Hurlburt Field, Florida where I learned some of the fine points of controlling air support.

In the summer of 1987, I reported in to the 1st Brigade, 25th Infantry Division. Rather than be assigned to a battalion, I was immediately slotted as the brigade assistant operations officer (AS3), a position normally occupied by a captain with prior command experience. Almost immediately I was handed responsibility for planning a battalion external evaluation (EXEVAL). Set for less than six months away, my boss, Maj. Lloyd Mills, later wrote on my evaluation, "Capt. Monger was the primary architect of the most complex joint training exercise and external evaluation of an infantry battalion ever attempted by this brigade. He coordinated resources from division, WESTCOM, Air Force, Navy and Marine Corps and supervised all aspects of the training exercise to include simultaneous operations on three islands." Mills was a phenomenal officer, and I was stunned to learn that in 1995 he was killed during night live fire training as a battalion commander in the 101st Airborne Division when a stray round struck him.

The reason I point this out is there is a common misconception among people without military experience, that soldiers merely follow orders and do what they are told. As a young captain, I was expected to take the concept and figure out operationally and logistically what was required for it to be conducted successfully. The traits young leaders in the Army rapidly develop can be wonderful assets for a civilian business. The veteran may not have the degree or certification listed in the job ad but is absolutely capable of performing satisfactorily or even exceeding requirements.

Following the Second World War, the United States retained the right to inspect the Japanese military to ensure it was not developing offensive capabilities—thus the title, Japanese Ground Self Defense Forces (JGSDF). I was picked to lead a small team of five (myself, a 2nd Lt., and three NCOs), to spend three weeks embedded with the Japanese 8th Infantry Regiment, an experience I will forever treasure.

Located on the northern shore of Tottori Prefecture, Japan in the town of Yonago, is the 8th Regiment Headquarters for the Japan Ground Self Defense Force. We had been warned by previous inspectors that the Japanese love to exchange trinkets—especially patches and unit crests—so we loaded up our bags with things to trade. When we arrived and met our hosts, we immediately pulled out a few items for them. They looked crestfallen, because the previous Americans hadn't been prepared to trade items, so the Japanese had put the word out internally not to embarrass the Americans by offering gifts. We had a laugh, and soon there was a line of Japanese soldiers outside our room, wanting to meet the Americans and trade a patch.

We were treated like royalty for the few weeks we were there and rapidly made friends with our Japanese counterparts. Highly professional soldiers, they were eager to show us their capabilities and acted highly impressed when I picked up one of their pistols and disassembled it to examine the inner workings. Toward the end of our tour, my translator, Lt. Miyamoto, told us the following day we would go for a hike to an ancient Buddha shrine. I asked where and he pointed to some small hills a few miles away. I gave a heads up to our team and we made sure to put extra shine on our Corcoran jump boots that night—we had to make a good impression!

The next day we loaded buses and drove right past those small hills, to the base of Mt. Daisen; according to a Daisen tour guide, "The third greatest summit in Japan behind Mt. Fuji and Mt. Yari." Standing nearly 6,000 feet in elevation, that mountain would be our "hike." Along with a company of Japanese infantrymen, we hiked a well-worn trail for about six hours. At the summit, we visited a thousand-year old Buddhist shrine, had our photo taken and began the descent back down. Along the way I met and had a photo taken with the commanding general of the Japanese Air Force, also out to climb the mountain with his staff. As we neared the bottom of the mountain, my knees were screaming with pain from the jarring impact of each step as Corcoran jump boots are made for show, not for hiking. I decided to jog to the front of the column of about fifty soldiers, and each one I passed picked up the pace along with me. Rather than a moment of relief for my knees, the descent turned into an all-out foot race to the bottom of the trail. We hit the end, sweating, smiling, shaking hands and slapping each other on the back—it was awesome.

The night before we left, the Regiment's officers gathered for a formal social with LT Wenzel and I. It had been a fascinating three weeks, learning the strange customs of bathing from a large soup bowl, toilets that were holes in the floor, eating with hashi (chopsticks) and learning new food.

About halfway through our stay, LT Miyamoto asked if we had any food requests, and we asked for burgers and fries. The next day we were treated to beef patties and crinkle-cut fries! When we picked up the fries, however, they were still frozen. Some things get lost in the translation. Their effort to please us, however, did not go unappreciated.

Back to the officer's social. It was an evening of Japanese beer, saki, and the best sushi I've ever had. By the end of the evening, all the badges and insignia from my uniform had been traded with Japanese officers, and I had their badges and insignia pinned on my uniform. The Japanese commander, Col. Michiya Teramoto, selected a very unique looking piece of sushi from the table. Not to be outdone, I reached for the same, only to have LT Miyamoto stop me, shaking his head in discouragement. I ignored him and popped the sushi in my mouth, and suddenly the entire room erupted in a cheer. I had unknowingly just eaten fugu, which is pufferfish. Improperly prepared, fugu can be lethally poisonous!

After the New Year, 1988, the brigade commander Col. Robert H. Wood, called me into his office and offered me command of B Company, 5th Battalion 14th Infantry. There was one kicker: I had to attend and graduate from Air Assault School before I could assume command — and by the way, the class starts Monday. No pressure.

The 25th Infantry Division had "coded" skills for certain jobs. An infantry company commander's job was coded both Ranger and Air Assault and would not be allowed to assume command unless he met both qualifications. According to www.goarmy.com, "U.S. Army Air Assault School is a ten-day course designed to prepare Soldiers for insertion, evacuation, and pathfinder missions that call for the use of multipurpose transportation and assault helicopters. Air Assault training focuses on the mastery of rappelling techniques and sling load procedures, skills that involve intense concentration and a commitment to safety and preparation."

Air Assault school wasn't especially physically demanding, although it did require completing an obstacle course and 12-mile rucksack march prior to graduation. There's always the possibility of injury when ropes and helicopters are involved, but I was eager to become an infantry company commander, so I immediately said, "Yes!" Then I had to go home and tell Jana, who was about eight months pregnant that I was about to take on one of the most demanding jobs in the US Army.

Thankfully, I sailed through Air Assault School, passing all the technical and physical qualifications easily. The week following graduation was consumed with the inventories and briefings required by regulation for an officer taking command of an organization. Then on Friday, February 26, I assumed command. After the ceremony, I met with the platoon leaders and platoon sergeants to tell them about my philosophy of command and to give them an opportunity to ask me questions. I received skeptical looks when I told them that the non-commissioned officers would conduct training—the previous commander had held a tight rein over everything that happened in the company, and it would take some encouragement and training to change the way they did business.

I would be helped in this by the new company 1st Sgt., Jimmy Akuna. Jimmy had joined the company several months before my change of command, and he was a highly respected figure—the epitome of professionalism and physical conditioning, and he had a commanding presence. Jimmy had served on a Ranger long-range reconnaissance patrol team in Vietnam and had been one of the original platoon sergeants when the 3rd Ranger Battalion was reactivated in 1984.

I talked my leaders through the "Four C's" of Leadership: Candor, Competence, Courage and Commitment. I'd read an article in one of the military professional journals and adopted that for my command philosophy, and it's been part of my thinking ever since.

The Four C's of Leadership

Candor: I encouraged them to feel confident telling me the honest truth even if it was unpleasant or embarrassing. Bad news doesn't get better with time; the sooner we identify an issue we can work to correct it. I promised not to "shoot the messenger" and asked them to do the same. I saw many toxic leaders while I was in the military (and even after). Exploding in anger or retaliating against someone who has identified a fault or deficiency only serves to make people cover up those deficiencies. I also pledged that I would not hide the honest truth from them, either. If I recall, I used the term "brutally honest." Let's not get hurt feelings, let's identify and fix things.

Competence: Mastering your profession means more than merely putting on a uniform and showing up for formation. It means studying the ancient as well as modern masters of war: Sun Tzu, Patton, Rommel, and others. It means being able to accomplish anything you expect your subordinates to do. Leaders, by definition, lead. I encouraged them to also study philosophy and psychology, to help them learn that expert leadership is getting someone to believe it was their idea to accomplish something, and the professional gives credit; he doesn't take it.

Courage: You must possess the courage to lead others into situations where you might be wounded or killed. We do very dangerous things in the Army, even during training. You must also possess the courage to be candid, and to accept responsibility for the things you need to improve.

Commitment: Being a leader in the military is your number one priority. You may get a call during your kid's birthday party that requires you to put on the uniform and report to work.

The very lives of your soldiers, and on a larger scale the freedoms our nation enjoys, rely on your total focus and unwavering commitment. If you can't do that, then you need to change professions.

Two days later, early Sunday morning, the phone rang at my quarters. It was an alert! I was to report to the unit immediately and prepare to deploy. To where and for how long was to be determined.

Turned out a training area at Fort Lewis, Washington was in terrible danger and required rescuing. The most danger I would face would be telling my nine-month pregnant wife I was leaving the islands and wasn't exactly sure when I'd return! This was the first of many such conversations. Later when I served in the 1st Ranger Battalion, we had a secret code. If I didn't come home from work she was to look for my parked truck. If there was a note inside that I'd gone for chocolate that meant I was deployed and would be back when I came back.

This particular alert was an EDRE (Emergency Deployment Readiness Exercise), and it was only my company that had been selected for deployment (and evaluation). Being in command for a mere two days, I had no idea of the level of training or knowledge of the team, but the company was a COHORT (Cohesion Operational Readiness and Training soldier replacement system) company, meaning the bulk of the company had gone through basic training together as one unit, with the company leadership picking them up from basic training and accompanying them as they formed the unit in Hawaii.

Theoretically, a COHORT unit will be more cohesive and better trained as they all start from the same point and grow professionally together.

While I might not have known the capabilities of the company, I was very familiar with the training area to which we had been deployed at Fort Lewis, for I'd spent three years there—although running around in tracked vehicles and HMMWVs—and now as a light fighter I'd be walking everywhere.

We flew on a C-141 aircraft to McChord Air Force Base in Washington State and unloaded near a gate that allowed direct access to a wooded training area. The company moved into a large perimeter, and in the dark, I called the platoon leaders to me. To avoid giving our positions away to any potential enemy, we laid on the ground in a circle, a rain poncho covering our heads in the center as we used small flashlights to examine our map. I had been given a FRAGO (fragmentary order) which gave bare details for our mission. In this case, it was to conduct a movement to contact, meaning we would walk through the woods until we bumped into the enemy, at which time we would assess the situation and respond.

Being the new guy and unfamiliar with our SOPs (Standard Operating Procedures), I asked the four Lieutenants facing me how the company usually left an assembly area. They looked back blankly, the light from the flashlights creating dark shadows on their faces. One responded that we really didn't have an SOP for that. Surprised, for this unit had been together for a year and a half, I told them how we would leave the assembly area and gave an order of movement. A few moments later, we were moving past 1SG Akuna and I was comforted to see that he was counting each soldier out of the assembly area. Think of this process as a collapsing balloon with air leaving the nozzle. As the air (the soldiers) flowed past the nozzle (a choke point), he touched each man. As the last one departed he recorded the total number, ensured it matched the number of soldiers we brought, and repeated this process every time we entered or left an assembly area thus ensuring we didn't leave anyone behind.

As we moved through the woods, I was pleased to see that the soldiers kept a good interval, spreading out into squad wedge formations. After about a klick (a klick is a kilometer), I knew from experience that we were nearing a road cutting through the area—a danger area and one ripe with ambush potential. Army doctrine requires certain procedures upon encountering a danger area that decrease the potential for ambush and increase security, but we weren't doing any of these things. I called a halt to the moving formation (over a hundred soldiers) and radioed the platoon leader up front. When he answered, I told him there was a road directly ahead, and he responded that he was looking at it. Once again, I called the lieutenants to me, and in the dark I asked them to tell me the company SOP to cross a danger area. I got the same response as before—there was none. It was going to be a long night.

The exercise culminated the next morning with a company attack on the MOUT (Military Operations Urban Terrain) site, called Regensburg. Regensburg mimicked a German village, with about twenty structures ranging from homes and small shops and even a small church. Once this was done, we conducted an AAR (After Action Review) and learned from the battalion operations officer, also called the S-3 that we would not be able to fly back to Hawaii for several days due to aircraft availability. I had the unique opportunity to on-the-spot plan and conduct any training I saw fit, and I was given a pallet of ammunition, access to ranges and use of Regensburg. Talk about a kid in a candy store!

The training areas in Hawaii did not contain anything remotely like a Regensburg, so that was the first opportunity I chose. Calling the company leadership in, I asked them for input on the training they would prefer and I told them the sergeants would be conducting the training. Based on their input, we devised a round-robin training session and identified sergeants that had prior training or experience in breaching, clearing, and other techniques essential for fighting in an urban environment. After a period of rest, we began training using a time schedule where each squad would work on a skill, then rotate to another station with rest periods for chow and water.

Midway through the morning training, my RTO (Radio-Telephone Operator) handed me the hand mic. The Commanding General (CG) was on his way to take a look at my training! Taking the radio from my RTO, I told my company executive officer to take charge; I was going to walk down to the gate to greet the CG. I'd call him once he arrived to let him know we were moments away. I headed to the gate, pleased at how the morning was going and looking forward to showing off my new command.

The general arrived, I saluted and reported to him and called my XO and hopped in the general's jeep. We rounded the corner to see … absolute chaos.

It looked like the Keystone Cops in fatigues. There was a squad of soldiers holding a ladder running one way, another squad running another. The general and I stood there dumbfounded. After a moment, he turned, patted me on the shoulder and remarked, "Well, you didn't have time to plan training," turned and left. I had no words, merely saluted.

I called my XO over to me. "What was that?" I demanded. He responded, "When you called to tell me that you and the CG were coming we were on break between rotations. Our old commander said to never let a general see soldiers just sitting around, so I told everyone to *do* something!"

We lost a huge opportunity to showcase the skill of our sergeants conducting training, and ended up looking like we were borderline incompetent. I'm lucky I didn't get relieved just a week into my company command. Now, it's a funny story but it is indicative of a too-common attitude (at least at that time) among some leaders in the military for form over substance. I don't blame them for this, I blame the former commander. My XO learned that would not be the way we did things going forward and he was eager to learn. He retired years later as a Lt. Col. after a successful career.

I returned to Hawaii intent on building this company into an effective fighting force, able to take on any challenge and win. I sat down at my desk, disassembled my .45 pistol and called home to let Jana know I was back and would be home in a few hours, once all the sensitive and valuable equipment had been accounted for, cleaned and turned in. She told me "you're coming home *now*, mister. We're on the way to the hospital." About twenty-four hours later Michelle came into the world.

Five months into my command, 5-14 Infantry got a new battalion commander, Lt. Col. James M. Dubik. Dubik ultimately retired as a three-star general and is now a member of the Ranger Hall of Fame. Dubik had served as a company commander in the 2nd Ranger Battalion and as the battalion XO for the 1st Ranger Battalion. He had taught philosophy at West Point and became a valuable mentor to me. His influence changed the course of my military career, and of my life. Under his leadership we took the battalion through a Joint Readiness Training Center (JRTC) rotation, considered the Super Bowl for military training. The officers and non-commissioned officers, and the soldiers of B Company (the Bushmasters) gelled and became, in my opinion, the best light infantry company in the United States Army at that time. As I changed command, my brigade commander praised me on my evaluation, stating, "Capt. Monger has been a real work-horse as commander of a rifle company in this brigade. He was carefully picked for this specific command to raise a mediocre company to a higher level of achievement. He has exceeded all expectations."

It was during my tenure as a company commander that I learned and truly grasped the methodology behind the Army's training system, and I became very good at it. I'll boil it down to the basics, and use one rifle company collective task as an example. I'm doing this from memory, not pulling it from a field manual, so forgive any incorrect terminology or lapses in detail.

For a rifle company to conduct an ambush, it must be able to shoot, move, and communicate. It must be able to emplace rifle platoons in security positions, support by fire positions and assault positions. These become broken down into subordinate unit collective tasks. A platoon in a security position must be able to emplace squads to cover sectors of fire, emplace hasty minefields, and communicate with all elements.

At the squad level you focus on individual tasks. A soldier must be able to properly employ his weapon. He must be able to draw a sector sketch, and a number of other individual tasks. Squad leaders are expected to keep a notebook with the status of each of their soldier's abilities to successfully accomplish their individual tasks. I blocked off time in the training schedule several times a week for squad leaders to be able to work with their individual soldiers on improving their performance.

Theoretically, if soldiers are trained to standard in all the individual tasks required for them to participate in each of the larger collective unit tasks, then the unit doesn't have to train as a whole on those collective tasks; it merely needs to train the leaders on how to employ the individuals and smaller units.

Each unit commander identified a Mission Essential Task List (METL). These were the items that the unit absolutely must accomplish to win on the battlefield, and each of those

METL tasks was rated as T (Trained), P (Needs Practice), or U (Untrained).

A rifle company that had never practiced an attack might be able to give itself a P rather than a U, because the subordinate units had each trained to standard on their own piece of the attack.

Once all these critical collective and individual tasks are identified and evaluated, it's a relatively simple task to develop a training plan to raise the U's to P's and the P's to T's.

My experience with the Army's training system helped me develop the framework behind GallantFew's transition system, a five-point approach to individual growth. I'll discuss this in Chapter 11.

We got the opportunity to find out how well we had accomplished our training. We were selected to participate in a Joint Readiness Training Center (JRTC) rotation. Considered the "super bowl" of peacetime training, it could also be a bet-your-future rank proposition. If you did well, that success would carry you forward. Do poorly and chances of attaining high rank would diminish.

JRTC is infamous for turning the most competent soldiers into sleep-deprived zombies, and the funniest thing to happen to me as a company commander occurred there, but it also proved to me my command had been successful.

The next-to-last mission of our ten-day rotation was a defense in depth, an arrangement of units that allowed for some maneuvering and areas to trap the enemy into kill zones, rather than a static and straight defensive line. Before we moved into position, however, my company had been selected for "the" plum mission of the rotation, an out-of-sector air assault, meaning we would board helicopters to do a sexy raid outside of the maneuver training area. Due to weather, though, we ended up doing a truck-assault, with an all-night movement to contact, to try and locate the enemy target. Finally, the middle of the next day, we found and destroyed it, but by then all of us were walking zombies, and many of us were simulated casualties.

To stress the unit and simulate combat, we wore MILES gear—lasers on our weapons and receivers on our gear. A blank round generated a small pressure wave that in turn fired the laser. If that laser beam contacted a receiver, the gear sounded an obnoxious chirp. The evaluators, called OC's or Observer-Controllers, carried decks of casualty cards. They randomly dealt them out and the card dictated the nature of your injury—a sucking chest wound or a mere scrape. Besides the fatigue of the mission now we had to treat and evacuate our casualties.

I'm 5'9" if I stretch, and weighed about 170 pounds then, plus another 50 pounds of gear, weapon, and ammo. A squad leader next to me was a casualty and I put him over my shoulder in a fireman's carry. He easily weighed 200 pounds, not including his gear—he was a big guy—and as I stepped I hit a depression in the ground, causing my lumbar to flex forward hard. I never let anyone in the unit know about that injury, but I did see the docs once I returned to Hawaii. They put me on a regime of traction and Motrin but I refused to miss work. Years later I had surgery to repair a bulging L5-S1 disc and to this day I have continued pain and sciatica, now developed into degenerative disc disease. Fortunately, it was recorded in my medical records and the VA recognizes it as a disability incurred on active duty.

Too many soldiers "suck it up and drive on" and don't report injuries, with the result that the VA refuses to recognize, treat, and compensate soldiers who have put their bodies through the most rigorous and demanding training, tearing up joints and ligaments. I've since also had a total hip replacement, but I never reported any of the discomfort from my hip and I continue to appeal my claim with the VA, who has not recognized it as a service-connected condition.

Carrying a "light" load at JRTC

Following the mission, I was called straight to the battalion command post for new mission orders, this for the defense. Once the briefing was complete, Lt. Col. Dubik and all four company commanders boarded a Blackhawk helicopter to conduct an aerial reconnaissance of the defense sector. The entire flight was conducted nap-of-the-earth, meaning the helicopter hugged the contours of the ground while flying as fast as possible to avoid being shot down. The enemy (called OPFOR for Opposing Force) had us in their sights and the pilots knew it. My entire experience of the flight was seeing blue sky and green trees, blue then green, blue then green, and by the time we landed, I hadn't seen a single thing that could help me plan this mission.

We landed and I learned that while I was being briefed, the company had moved into and occupied their defensive positions. I was in such a fog then that today I don't recall whether my RTO and I had to walk several klicks to link up with the company or we were given a ride. I do recall lying on the ground with my platoon leaders, poncho covering up our heads once again as I relayed our mission to them. The stench that comes from six combat soldiers who have been repeatedly covered in sweat, dried, then covered in sweat combined with the smell of Copenhagen, stale cigarettes, and confined under a rubber poncho in a face-to-face circle, is unlike any other.

I woke up perhaps an hour or two later, still laying under the poncho but totally alone. My first thought was to find my rifle; there it was right next to me. I sat up, pulling the poncho off and felt the coldness of an Arkansas night in March hit my face. I was completely alone. I pulled on my night vision goggles and in the green fog could see a faint glow from several hundred meters away and, listening carefully, I caught the occasional clang of an entrenching tool hitting rock. Making sure I didn't leave anything on the ground, I carefully made my way to the sound. A moment later I walked up on a pair of B Company soldiers digging a fighting position. They greeted me and we chatted for a moment then I asked them, "Where was the company command post?" They pointed it out, most likely wondering if it was a test. Finding the XO a few moments later, I asked him what the hell were they thinking and why had they just left me there?

He responded that I had fallen asleep in mid-sentence and they knew I was tired, I was in a safe enough place, and they knew what they needed to do so they decided to let me get some rest. That to me was confirmation that I had turned this company into a competent, confident fighting force. What a change from a year earlier!

Following the JRTC rotation, the COHORT company's three-year cycle concluded. Some of the soldiers had been promoted to leadership positions—team and squad leaders—but most of the soldiers either left the Army or re-enlisted and accepted new assignments elsewhere. Now I had the privilege of accepting an entire new COHORT company and the challenge of training them. I was eager to turn this group into an effective force that I would be proud to hand to the next commander, and together with Jimmy Akuna, we had developed superb leaders that made this a reality.

Midway through my command tour, Lt. Col. Dubik called me into his office and suggested that I put in a packet for assignment to the 75th Ranger Regiment. Unlike any other unit in the Army, an officer must apply, and as a leader he must have performed the same job in the regular army first. Dubik went on to tell me how to put together a packet, suggesting that I contact all my previous battalion commanders to request a letter of recommendation. Doing some detective work, I tracked all of them down — receiving four sterling letters, each on official, Department of the Army letterhead. Surprising to me, my very first battalion commander (who went on to command the 3rd Ranger Battalion) wrote that had I been available then, he would have actively pursued my assignment to that battalion. Hale had been my commander when I received my letter of reprimand.

Completed packet in hand, I carried it into Dubik's office and said to him, "Sir, my packet is complete. All I need is your recommendation." He reached into the top drawer of his desk, pulled out his personal stationary, with a beautiful golden dragon embossed at the top (the 14th Infantry is the "Golden Dragons") and using a Sharpie, wrote, "Buck, hire this guy. Jim." He tore it from the pad and handed it to me. "You want me to send this?" I asked. He replied, "It's better than any other letter in that packet."

Col. Buck Kernan was then the Commander of the 75th Ranger Regiment, and he and Dubik had served together in the 2nd Ranger Battalion. Off went the packet. Somewhere around this time I met Maj. Ken Stauss, friends with Dubik, and pegged to be a future Ranger battalion commander. I didn't know it then, but Ken would become my friend and my boss, and his life and death would impact me in ways unimaginable.

For over a year and a half and with two distinct COHORT companies, I had held the best job in the United States military—that of infantry company commander. It is the first level in the Army that is called a "command" position and it is the last level that an officer has daily, intimate interaction with junior enlisted soldiers. As company commander, I was the one responsible, should we go into combat, for having prepared the unit to be successful and defeat the enemy. Along the way, I became mother and father, pastor, and judge to them. I had to break the news to a soldier that his mother had tragically died. I had to rout out and punish drug use, a cancer in the military and readily available to young soldiers with money. I put soldiers into confinement for misbehavior. I led them by personal example and never asked or expected them to do anything that I wasn't prepared and able to do myself.

Even Jana became part of the unit. She shepherded soldiers' wives, even taking the place of a squeamish husband in the delivery room. She would be recognized in the 25th Infantry Division Spouse of the Year ceremonies and remains highly proud (and rightly so) of the coin the division commander gave her.

Chapter 5

I Loved These Soldiers

I am not ashamed to say that I loved these soldiers.

I like to believe that B Company, 5th Battalion 14th Infantry Regiment was the best company in the US Army. I was far from a perfect commander—lest anyone think that I proverbially walked on water. I made mistakes, but I believe I learned from those mistakes and I did my best to be a good commander. I was blessed to have professionals surrounding me, men who constantly strove for excellence. They deserve the credit for our successes.

One of the terrible burdens of command is dealing with the soldier that does not perform—especially a soldier who is in a leadership position. B Company had a squad leader who was mediocre, and he had an alcohol problem. I learned about the alcohol when he missed formation. I proceeded to do what I see many commanders do today: I used his alcohol problem to destroy his military career.

As a commander charged with preparing a unit for battle, I did not want a mediocre leader in my unit. I saw the potential for his poor leadership to cause a mission to fail, or a soldier to be maimed or killed, and I saw the misbehaviors his alcohol abuse created as a rapid path to remove this problem.

I brought him before the battalion commander for a field grade Article 15, because the battalion commander could inflict greater punishment than could I. The battalion commander did inflict severe punishment, but he suspended the worse portion of it, under the condition that if he misbehaved again the punishment would automatically kick in.

It didn't take long. A week or two went by and he missed a formation. I asked, and the battalion commander lifted the suspension, and the staff sergeant became a sergeant.

As company commander, I had the authority to reduce him in rank to a sergeant. He missed another formation and became a specialist.

I made no attempt to diagnose the reason why he drank, because to me he did not demonstrate the leadership potential that would help me improve the unit. It was a brutal but effective way to remove a mediocre performer from our midst.

Now it wasn't all on me; he had made decisions that led to the punishments. I remember asking him when I reduced him in rank from sergeant if he wanted to leave the Army as a private, because if he continued his behavior that was exactly what was going to happen.

I wonder where he is now, and how his life turned out. I hope and pray he found help and peace.

I see veterans like him often. While in 1989 we were not a nation at war; now we have men and women who are summarily dismissed from the military, some with other than honorable discharges (OTH), for offenses with drugs and/or alcohol. Many of them I am convinced have post-traumatic stress and traumatic brain injuries, and they have tried to conceal it from their superiors for fear of losing their leadership position or assignment to an elite unit. Any perceived weakness or potential failure to complete the mission especially in the special operations environment is met with swift and sudden removal.

It's a double-edged sword. Report symptoms of Post-Traumatic Stress (PTS) or Traumatic Brain Injury (TBI) and lose your leadership position. Use drugs and/or alcohol as self-medication and lose your leadership position, and most likely your military career. A veteran with an OTH discharge has an incredibly difficult time gaining employment. A veteran with a dishonorable discharge does not qualify for care from the VA.

There is a very fine line between crushing an underperforming, substance-abusing soldier and identifying a legitimate cause for a soldier who has incorrectly sought solace from substances. When a person has otherwise served honorably, I charge the chain of command to get that person help, not to crush their soul. I don't have the answers to this, because as a former commander I totally understand mission focus and the drive to shed anything or anyone that detracts from that—but we shouldn't destroy the rest of a person's life as we remove them unless they engage in real criminal activity.

My saddest day in the Army (not counting memorial services) was the day I relinquished command of the

Bushmasters. A change of command ceremony is a formal event. The unit flag is a guidon—a two-pronged pennant in the color of the branch (blue for Infantry) with the unit designation in white lettering. It is carried by the best soldier of the unit. The First Sergeant, incoming and outgoing commanders, and battalion commander stand in a square, facing each other. The First Sergeant retrieves the colors, hands them to the outgoing company commander, who passes them to the battalion commander, symbolically giving up his hold on the unit as he lets go of the guidon. It took a conscious effort to uncurl my fingers from the wooden guidon staff, but before I did, Lt. Col. Dubik in a low voice that only I could hear, said to me that the nation will never know or recognize the contribution I had made to our country. He told me he did, and he thanked me.

The next day I found myself right back in the same chair, behind the same desk in the brigade operations section, in my old job as brigade assistant 3. I was back on the battalion external evaluation train, but now with a new brigade commander, Col. David Ohle. Ohle also retired as a three-star general and became a valuable mentor in the short time we served together. He fully supported my request, seeking assignment to the Ranger Regiment, and although a general at division HQ denied my request for assignment, Ohle wrote me a letter of support, and in his letter to the US Army Personnel Command at the Pentagon he called me the best captain he had in his Brigade and that he planned to give me a second command, that of the brigade headquarters company. It was one of the defining turning points in my life, with more to come. Had I stayed and taken a second command there and done well at it, I might have stayed in the Army; I might still be there today. GallantFew and the work we are doing to help veterans might never have happened.

In December of 1989 I received an official welcome and acceptance letter from the 75th Ranger Regiment. The very next day the Regiment conducted a combat parachute assault into Panama for Operation Just Cause.

When I left Hawaii, I attended the sixteen week Combined Arms and Staff Services Course, known as CAS³ for short, and pronounced "kass-cubed." CAS³ was graduate level PowerPoint, where captains were trained and practiced on how to properly produce every conceivable type of presentation. I spent my spare time carrying a heavy rucksack around the army airfield at Fort Leavenworth, Kansas to make sure I'd be ready for the Rangers.

Fort Leavenworth was only a few hours away from my family in Wichita, and one weekend I drove home to visit. My grandparents had arranged a big get-together at their home, and my Great-Uncle Roy Jones, 94 years old and a veteran of World War I, drove in from the ranch he managed a few hours outside of Wichita. I had not spent much time with him, and late in the afternoon he and I sat on a covered patio, no one else around.

Uncle Roy began to tell me about his experiences in the Great War where he had served as a Corporal and a squad leader in Company G, 353rd Infantry, 89th Division. He told me stories of buddies that had died, of sending men out on patrol that didn't return, and as he talked it was as if he was transported back to that time. He talked of hearing his friend scream in agony after getting hit by an artillery fragment, and how he counted the seconds between the incoming bursts. He waited for the pause, then ran to his friend and dragged him back to his foxhole. Amazed, I asked if he had been decorated for valor and he recoiled as if I had slapped him in the face. "I didn't do anything brave," he said.

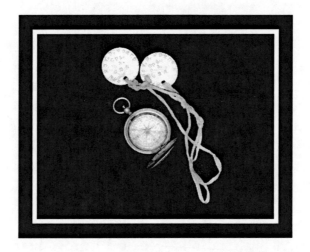

My Great-Uncle Roy's compass and dog tags
Roy Jones served as a corporal and infantry squad leader
in the 89th Division in World War I.

The next day I went back to my grandmother's home, and while talking with her I mentioned her brother, and how brave it was that he had saved his buddy. She almost dropped the plate she was holding, and she looked at me in shock. "What?" She exclaimed, and told me he had never told anyone else in the family those stories.

I didn't realize it at the time, but now that I was an infantryman, and although I hadn't experienced combat, a special relationship now existed between the two of us. Now that we both were infantrymen, that made it okay to share his stories. I believe it helped him to share, too.

I was so blessed by that conversation and the time we had together, for he passed away the following year. I inherited the compass he carried in France, along with his dog tags, canteen cup, mess kit, and a few other mementos.

In July of 1990, I reported in to the Ranger Orientation Program (ROP for short, pronounced "rope"). ROP was a two-week trial period for officers and senior non-commissioned officers reporting to the Regiment. We were required to recite the Ranger Creed, we administered physical fitness tests including a timed twelve-mile road march, and practiced sticking each other with needles so we would be proficient at giving IVs. This was the first ROP course conducted after the Regiment returned from combat.

I wanted to test my fitness against some Rangers, so I showed up at the Ranger Indoctrination Program (RIP) area early the Friday morning before my ROP class began. RIP was physically and brutally demanding with a high attrition rate. I picked out the nearest squad forming up and introduced myself to the RIP instructor. I stood out like a sore thumb, as I had to wear the old Army gray PT uniform, a heavy cotton tee with ARMY across the chest and matching bulky shorts. Once a Ranger completed RIP or ROP, he was then authorized to wear the Regimental PT uniform, a tight fitting black tee with the unit scroll on the left breast and the silky running shorts known as "Ranger Panties." I told the sergeant I was reporting in to ROP Monday morning and might I run PT with his squad? He replied, "I don't think that's a good idea, sir. We don't want these soldiers to see an officer fall out of a run."

I was nearly ten years older than this sergeant, but I could run. The challenge was on and I still feel sorry for the kids in that squad. The sergeant took us out at a blistering pace without telling us how far we were going. As we passed the five-mile mark, my watch showed we were averaging a six-and-a-half-minute pace, and we were now headed up a hill known affectionately at Fort Benning as "Heart Attack Hill." I was well familiar with the hill, having run it many times during my assignments there at the Infantry Officers Basic and Advanced Courses. We started the run with ten soldiers, and by the time we got to heart attack hill there were half that number, with me the last man in the column. Two of the kids face-planted going up that hill, and I had to hurdle over both of them. After we reached the top, the sergeant showed no signs of slowing and I sensed he would go until he was the only one left, however long that took. So, I ran up next to him and told him I was breaking off, and that I planned on going to a longer route. The reality is that by then I thought I was going to have a heart attack. He and the survivors went right, I went left, and we both preserved our pride.

Two weeks later and Col. Kernan was congratulating me on successful completion of ROP and I was authorized to wear the coveted black beret of the Ranger Regiment. I reported in to the 1st Ranger Battalion stationed at Hunter Army Airfield in Savannah, Georgia. I was both conspicuous and self-conscious as I immediately felt like an outsider, being the only officer in the unit that didn't have a Combat Infantryman's Badge (same as my EIB but with a silver wreath encircling the rifle), a "mustard stain" — a small gold star on the jump wings indicating a combat jump, and right shoulder combat scroll patch. It looked like that would soon change, as a week after I got there, Iraq invaded Kuwait and the first gulf war began.

For years, I had said a simple prayer every night, "God, let there be peace. But if there must be war, then send me first." God had other plans for me, because I never seemed to be in the "right" place and at the "right" time. First, I was in Ranger School during Grenada, then I had orders assigning me to the Rangers but watched it on CNN from Hawaii, and now with the entire United States military going to war, the Ranger Regiment stayed behind.

Chapter 6

Flesh-Peddling

It had been seven years since the last time I jumped from an airplane in flight. I'd jumped five times in jump school and five more times in Ranger school. We were supposed to get several jumps at ROP but with the rest of the Army deploying to Saudi Arabia, there were no aircraft available. My airborne refresher consisted of sitting in a classroom reviewing the various things that could get a jumper killed.

Now I'm sitting on the tarmac at Hunter Army Airfield, about to make the first night jump of my military career. No Ranger wants to appear nervous, insecure or anything but totally in control—but I'm about to jump with a time-on-target of midnight out, of an aircraft with no lights on inside, over a drop zone with no lights on the ground. I could either "fake it" and risk whatever happened or I could swallow my pride and ask. I chose the latter, confiding in the major waiting next to me (and who had come to the Rangers from the 82nd Airborne and who had a lot of jumps) that this would be my first night jump. He told me not to worry, I'd be able to see more than one would think and to rely on my training—and he was exactly right.

The adrenaline was rushing through my veins as the jumpmaster went through the commands: "Outboard personnel stand up," "Hook up," and so on. Through the small portal window next to me, I could see lights below, illuminating the occasional street corner or parking lot, surreally floating by in the otherwise jet-black scene. In what would become a ritual to me, I recited the Lord's Prayer in my mind, and heard the command, "Ten minutes!" The Air Force crewman pulled open the doors on both sides of the aircraft and cool, fresh air flooded the bird. "Thirty Seconds!" The Ranger behind crowded me, pushing me forward as the stick of paratroopers surged towards the door. "Go!"

This jump was from a C-130. A prop plane, the C-130 has been the workhorse of the Air Force since Vietnam. It flies slower than the jet transports, and it requires a slightly different exit technique. With a jet, one slips out of the door almost like one would recline on a sofa and the slip stream does the rest. This plane, however, required a jumper to grab the edges of the door and leap forward, propelling himself away from the aircraft.

I had the best exit ever, no doubt due to the extra adrenaline. It was such a great exit that the opening shock pulled me smoothly horizontal to the ground and afforded me a majestic view of the aircraft, seemingly frozen in time and space, floating above me. Time sped back up and I swung under the canopy, now in near silence. Close your eyes, and that's about what I could see with my eyes open.

As I descended I could sense the earth coming; by golly, the major was right. Following my training, I kept my eyes looking to the horizon rather than down. When I sensed what I thought were tree tops, I dropped my rucksack (when rigged the ruck is tied upside down to the front of the jumper's legs and when released drops about 30 feet to give clearance to the jumper and "riding in the ruck" is a sure way to get hurt), make sure my feet and knees were together but relaxed and waiting for the impact.

Most of my landings were similar to sitting down in a chair, only a practical joker pulls away the chair. This was no different, but I landed in a grassy area, unhurt and intact. I released my canopy to remove the risk of being dragged by the wind, put my weapon into operation, gathered my gear and moved out to the assembly point.

Jumping in a Ranger battalion is different than in most airborne units, or so I learned from the other Rangers who had more experience jumping. Jumping at night, at low altitude, and with a hundred or more pounds of gear carries with it a high probability of injury. Injury in a Ranger battalion meant reassignment away from the unit — therefore Rangers only jumped frequently enough to retain proficiency (minimum of one jump every three months or lose the extra $150 a month hazardous duty jump pay), and when they did get injured, unless it was something requiring obvious care, they didn't report the injuries. I ultimately jumped 21 times with the 1st Ranger Battalion; all but three of them were at night and one was from a massive C-5 transport.

Capt. Monger, rigged for a jump in Panama

In those jumps, I fractured an ankle, hit hard enough to jar my skull, chipping and knocking a tooth loose, strained my low back, hit my head hard enough to see stars, and once landed on concrete splitting my elbow open to the bone. The fractured ankle is the only injury in my records, but only because I had an x-ray.

How does a Ranger deal with injuries? He hides them, he swallows massive amounts of Motrin, and takes preventive measures so the injuries don't reoccur. While today it's common to see soldiers wearing knee and shoulder protection, back then it wasn't part of the uniform — it was not authorized. So, we wore knee and elbow pads but we wore them under our uniforms, where no one could see them. Not very comfortable after you land when you are on an all-night movement, but it was worth the discomfort to minimize the risk of injury. I also wore a boxer's mouthpiece to prevent knocking out a tooth. I also visited the battalion Physician Assistant after duty hours, a legendary Ranger Warrant Officer named Doc Donovan. Doc would put me on a table and try and turn me into a pretzel in an attempt to release the cramping in my spine. To this day I can't stand in one place or sit in a hard chair for long without leaning against something or fidgeting around because of the cramping that builds in my back and the pain that runs down my legs. Despite surgery and a history of continuing pain management with the VA, my disability there is rated at ten percent, but only because of the injury I sustained as a company commander — the subsequent physical therapy was documented, and that provided the justification for the rating.

As the newest infantry captain in the battalion (and the only one without a combat patch), I was assigned to the S3/operations as the "Ranger Coordinator." In effect, it's a combination of local liaison and special projects that falls outside the scope of the traditional staff positions, such as the S-3 air (the officer that was responsible for coordinating all aircraft assets), the assistant S3 (the primary backup for the major who held the S3, or operations position). It's also the job for the new guy, so he gets the jobs no one else wants to do!

A battalion staff (at least at that time) is composed of five major sections. The S1 is the personnel section, responsible for evaluations, pay, mail, chaplain, and medical support. The S2 is the intelligence section, responsible for figuring out what the enemy is up to. The S3 plans training and combat operations and as such is the largest section. The S4 is the logistics section, handling all supplies, fuel, ammunition and transportation. The S5 was the civil-military affairs section, interfacing with local government and outside resources including media and public relations (today that section is designated the S9, the S5 is special projects). The primary staff officers are known by their section, for example the personnel officer is called the S1, and the primary staff officers report to the battalion executive officer (second in command and chief of staff) and then to the battalion commander.

I was given two major assignments. First, I was designated the rear detachment commander should the battalion deploy to what was then the beginnings of Desert Storm. To be the only officer in the battalion without combat experience and told that when everyone else went to war you'd be left behind was a crushing blow and told me I wasn't yet part of this team. The other assignment was to build the first live-fire CQB house. CQB stands for Close Quarter Battle. Only a few units in the army could practice clearing rooms and buildings using real bullets, and the 1st Ranger Battalion was soon to be one of them. I became the project manager, coordinating the materials and labor for what was a million-dollar project. Although I didn't have any construction experience, I had managed timelines and budgets before and this was right up my alley—a simple project.

Then I made what was probably a strategic error in my Ranger officer career. I knocked on the battalion commander's door and asked him to reconsider leaving me behind as the rear detachment commander. I did it for two reasons—I felt I needed to be in the fight and I thought it was the right thing to do. In retrospect, I realized it probably made me look self-serving and not a team player and I believe it created a negative impression in an officer that would come back to haunt me later. I intended just the opposite.

Ironically enough, the Ranger Regiment didn't deploy to either Desert Shield or Desert Storm. There were other threats in the world and we were the only major unit in the military that wasn't headed to the Middle East. Watching on CNN every other soldier (it seemed) in the world going to the desert, was frustrating. Imagine the Super Bowl where one of the champion teams was told they weren't allowed to play in case another important game came up and you might get a sense of what we felt.

In the fall of 1990, President Bush came to Fort Stewart. The post arranged a huge rally, with mostly families and those soldiers not deployed filling the sports stadium. Our battalion commander ordered up a slew of buses and the entire Ranger battalion (at that time the only unit in the Army to wear the coveted black beret) occupied a huge portion of the stands. I believe his intent was to let the President know we were here—ready to go. If that generated any questions as to why the Rangers were still stateside, I never heard about it.

I soon learned that Lt. Col. Ken Stauss was due in to take command. I also learned that I had been selected to be the new S1, a job in the Army I had never held before. I was totally comfortable in the operations environment — planning training and large exercises. This would be a learning experience from day one, as I in effect became the Human Resources Director for an 800-person business.

The Ranger battalion is unique in that Ranger battalion commanders (at least then) got to hire their own officers. In the rest of the Army, an officer lists his or her top three assignment choices and the Army decides where you will go. To be assigned to a Ranger battalion, that officer must apply, and must have done the job to which he is applying somewhere else in the regular Army, and it helps to have connections to get you considered. When Ken Stauss took command, he gave me a list of officers he wanted in the battalion and I went to work to get them to put in application packets.

Before Stauss arrived, however, we received orders to deploy a portion of the battalion to Desert Storm. Now as the S1, I stood on the airfield, counting noses and bidding goodbye to another group of Rangers headed off to war — again without me!

Remember my prayer?

> *"Lord, let there be peace on earth — but if there is to be war, then send me first so others don't have to go."*

My destiny would be different.

That small portion of the 1ˢᵗ Ranger Battalion deployed and became a part of the Regiment's history, and I am intensely proud of the role I played in getting them there and home again safely.

Ken Stauss was a breath of fresh air. He was a young Lt. Col., and had been the Regiment's Executive Officer during the Just Cause invasion of Panama. He always referred to it, in his Arkansas accent, as "Just Cuz." We immediately hit it off.

Lt. Col. Ken Stauss rigging for a jump in Panama

As a brand-new personnel officer, having the benefit of zero official training in the arts of Human Resources (in the Army often called, "flesh peddling"), I had to learn everything through trial by fire. Fortunately for me, the military uses checklists and standards for everything, and the staff sections that I led were each led by Rangers—so I often felt like the Colonel in the TV show *MASH*, asking for a document that a sergeant was already handing to me.

A key function of the S1 is to be the commander's administrative assistant, or "Adjutant". Ranger battalions are also unique in the Army in that battalion commanders have civilian administrative assistants and these professionals are the historical knowledge base for the unit. They stay for decades, while the officers rotate in and out every year or two. The 1st Ranger Battalion was blessed to have Shelia Dudley in that role while I was there and she retired in 2016, receiving a well-deserved US Army Special Operations Command Lifetime Achievement Award. Shelia often kept me out of hot water and I treasure her friendship to this day.

I soon realized I was the designated gatekeeper to the commander for all the other staff officers (except the battalion executive officer who was second in command). The first day on the job, one of the staff officers walked into my office and gave me a document, asking if I could get the commander's signature. I said sure, and walked to Ken's door, verified he wasn't in the middle of something and asked if he would sign a document. He said sure, looked at it and signed it. An hour or two later, another document, another trip to Ken's desk. Another hour, another document. As I turned to walk out Ken yelled, "S1! What are you doing to me?"

I turned back around and he met me in the middle of his office, ever animated as he spoke as much with his arms as with his mouth. He said, "Three folders. Red, yellow, and green. Red, I need to see today. Yellow, sometime this week.

Green, who cares if I ever look at it? Drop them on my desk first thing in the morning, I'll get the Red folder back to you before I go home; you'll get the Yellow folder at the end of the week."

"Roger that, sir," I replied, and went hunting for folders.

The next day a staff officer brought me a document. I asked him for the priority, and he said, "I need it now!"

I placed it in the Red folder and set it aside on my desk. He looked at me expectantly. "Aren't you going to go get it signed?"

The system worked beautifully. The staff officers had to be trained and I learned to assess how real the priority they assigned was, so that everything didn't go in the red folder. The result? I'd drop the red folder on his desk before the 6:00 a.m. formation (0600 in military speak), and he'd usually give it back to me by lunch, many times with the yellow folder action items also completed. This is a system I still use to organize my work load today.

The way that Ken "trained" me instantly elevated my respect for and desire to follow him. He didn't raise his voice, didn't make me feel stupid; he simply identified that I had a training deficiency and he trained me. This calm yet forceful leadership of requiring excellence without demeaning is a hallmark of every one of the Ranger officers I served with that went on to become General officers. It's a leadership style I use today, and when I work with veterans who are going through difficult situations I find this especially useful. Require excellence, expect it, and help others to achieve it.

Due to the demands of being the battalion personnel officer, I had developed a habit of coming in to work at 4:00 a.m. Such is life in a Ranger battalion—merely meeting the standard is not enough. You must daily work at a mental, physical, and emotional level that far exceeds any other assignment experienced in the Army.

After a couple of hours knocking out work, I'd jog a couple blocks over to the Headquarters Company area and get in formation with the other staff officers and sergeants, and we'd run our daily Physical Training (PT). Even officers aren't exempt from formation in a Ranger battalion. We'd do our exercises by section (me with my S1 crew) and then break down into speed groups for a run. I was quick, so I always ran with the fast group. We'd break into three groups—six-minute-per-mile pace, seven-minute and eight-or-slower pace.

Once a week the battalion executive officer held staff officer PT—often a gut-check of epic proportions. In keeping with the fluidity of a combat situation and to develop the ability to overcome unknown stressful situations, staff officer PT was a departure from the traditional task-conditions-standards that required leaders to tell soldiers exactly what training was about to occur, under what conditions and to which standard they would be held.

Example. Task: we will run five miles. Conditions: we will wear tee shirts, shorts, and running shoes. Standard: we will maintain an eight-minute-per-mile pace.

Staff officer PT was an unknown. The only person who knew what we were about to do was the major leading the PT formation. It is a true physical and mental test to start out on a quick run, in formation, not knowing if we were about to go two or ten miles.

Soon I learned to do reverse psychology on myself. When we approached the road intersection, where to turn around meant a five-mile run, to turn right meant a six-mile run, and to turn left meant a ten-mile run, I mentally told myself "we're turning left." I prepared myself for the most difficult option; then, when it happened, I wasn't caught hoping that we would turn around.

Sometimes we would turn left, go a short distance, then turn around and go back to the shorter route. You did not want to be the person who gave up and dropped out of the run, only to have everyone else abruptly turn around and head back towards you — yet even in that ultra-competitive environment, soldiers mentally sabotaged themselves. Usually, once they saw what was happening, they started running again, meaning to me they had the physical ability all along.

Without knowing what I was doing, I was putting the Burris Functional Emotional Fitness ™ techniques into practice, eliminating negative self-talk and replacing it with positive self-talk. More on that in Chapter 13, Functional and Intentional Fitness.

I remember looking at some of the older officers I'd served with in other places who huffed and puffed to keep an eight- or nine-minute pace. Then when I became forty, I had a much greater appreciation for the fantastic shape they were in!

In the summer of 1991, Regiment held Ranger Rendezvous. Every two years the Regiment changes command. Ranger veterans converge on Fort Benning for a week of fellowship, weapon displays, athletic competitions, and, of course, parades.

Perhaps you've seen a military parade or change-of-command ceremony. What you didn't see was the days of preparation leading up to the ceremony. Rehearsals are held starting with key leaders, meaning the officers and other key individuals responsible for making sure units get to the right place at the right time. In this case, as battalion personnel officer, I was a key leader. The staff officers stand in a line directly behind the battalion commander and in front of the battalion.

Columbus, Georgia in July is hot and muggy — borderline miserable — except we'd come from Savannah which is worse. Standing at alternating positions of attention and parade rest while the event organizers get their timing perfected, is tedium and a physical stamina test. Periodically, while the staff worked out bugs in the sequence, we would be called up to the front of Building 4 (now called McGinnis-Wickam Hall) to be told what we did right and what we did wrong, and we eagerly grabbed iced bottles of water to rehydrate.

Standing in the shade offered by the four-story building, I noticed an older gentleman standing nearby. I smiled and said something like, "Good afternoon, what brings you out here today?" to which he smiled and said simply, "My son is in the ceremony."

A moment later they called staff officers back to their positions, and we double-timed (meaning ran) back to our respective designated spots. Our exact spot was identified by a large soup can lid pegged to the ground. We learned quickly not to put a combat boot atop the metal lid, because after a few moments it felt like standing on fire. The lid reflected the sun and heat straight up through the sole of the jungle boot, and the steel shank in the boot amplified the heat.

As we took our positions, I heard over the loud speakers, "And now Lt. Gen. Grange will move to the center of the field," and the distinguished gentleman I had just met marched to the Regiment colors in the center of the field.

I could have died of embarrassment. Lt. Gen. Grange is a legend in the Ranger community — in fact in the military. The annual Best Ranger Competition, held since 1982 is, "The David E. Grange, Jr. Best Ranger Competition." Lt. Gen. Grange served in uniform for 41 years, first as an enlisted paratrooper in World War II, then as a platoon leader and company commander in the Korean War, then as a battalion and brigade commander in Vietnam. He is the epitome of a Ranger, and I should have known who he was. But when I asked him what brought him out today, his answer was simply, "My son is in the ceremony."

Twenty years later I had a chance to sit down with him at the Best Ranger Competition at Fort Benning. He laughed when I told him that story.

That is humble leadership. That is a man I would follow into hell.

A Ranger's life is never dull. In October, 1991 I learned just how dark a Panamanian jungle can be at night, and I looked longingly at luxury cruise liners passing by in the Panama Canal. The following month, I traveled to South Korea with the other Regimental personnel officers — the 75th Ranger Regiment S1 (a Maj.), and the 2nd and 3rd Ranger Battalion S1s. We spent a week traveling to military installations, making presentations, and encouraging officers and sergeants to apply for a Ranger assignment. I had the privilege of touring the DMZ, the line between North and South Korea created at the Korean War cease-fire. In Korea, the war hasn't yet ended, it's only paused.

At Panmunjom, the United States controls the southern side of the DMZ, North Korea the opposite. Straddling the border is a Quonset hut that North and South Korea used until 1991 for negotiations. Each side entered the building through a door on their side, and a conference table sat exactly on the 38th Parallel, the designated DMZ. Dividing the table exactly in half along that line was a microphone cord, linked to microphones that faced both to the north and south. Outside the building, American MPs, specially selected for their size and imposing presence stood at the two corners, with only one-half of their faces and bodies exposed to the north. On the opposite side, North Korean soldiers did exactly the same.

Every hour, control of the building changed from north to south. We were permitted to enter when our side had possession, and as the officer giving us our tour talked, he told us we could go on the other side of the table, "into North Korea." So, of course, we did—and I'm sure the North Korean soldiers were keenly aware of the Ranger scrolls on our uniforms. We stood near a window on the north side, and a tough-looking North Korean soldier stood at the window, two feet away and glared at us.

I had a disposable 35mm camera in my pocket, I pulled it out and tossed it to our guide and I have a priceless picture of a North Korean soldier peering over our shoulders while we stood on his side of the border.

1991: I'm on the north side of the DMZ.
That's a North Korean soldier looking at us.

On my return trip to the States, I ran into a Regiment liaison officer in the Seoul airport. He asked me "Are you going to get back in time?" To which I answered, "Back in time for what?" He said, "Your battalion's been alerted to deploy to Kuwait."

I got back in time. As I landed at the Savannah airport late at night, two of my sergeants were waiting. They drove me home, I grabbed my bag with the desert camouflage uniforms, and headed to battalion headquarters. We had been alerted to parachute into Kuwait.

At the end of the ground combat phase, the US and allies had withdrawn back to Kuwait and Saudi Arabia but the resulting situation was more like that with Korea — rather than a clean defeat and post-war rebuilding, now we had an existing border between two continuing adversaries. Saddam Hussein was rattling his saber again, and was moving troops south towards the border with Kuwait. The First Ranger Battalion was alerted to conduct a Show of Force to remind him of the consequences he faced should he decide to again invade.

This was a political dog and pony show, yet it was a deadly serious mission. We received our operations orders, drew live ammunition and headed to the aircraft to load up. By the time we landed in Madrid, Spain to change aircrews, I had traveled from Korea to Europe, the long way around in less than two days. I had a splitting headache and could barely walk. Fortunately, one of the medics noticed and gave me something that knocked me out for a couple of hours. I awoke refreshed and ready to go, in time to watch the aerial refueling tanker moving into position as we flew along the Saudi - Iraq border with fighter escort. We were making sure that Iraq knew we were there.

Besides the migraine medication the medic gave me, every Ranger on the mission was given Halcion to help them sleep on the flight. That same year the drug was banned in

Britain because of troubling side effects including extreme depression and suicide risk. So here we have an entire battalion of Rangers, armed to the teeth and more, jumping into a foreign country, taking medication that causes depression and in doses that were not documented in their medical records.

In the present day, there has been controversy over giving the drug Mefloquine to combat malaria. Turns out the drug's side effects mimic PTS and TBI and may damage the brain. Our soldiers were freely given it, however, with zero documentation in their medical records. No record of frequency or level of dosage. Try proving that to the VA.

http://www.militarytimes.com/story/military/2016/08/11/ malaria-drug-causes-permanent-brain-damage-case- study/88528568/

During the air mission brief we'd been told that due to the extended distance we were flying the aircraft, we would have only enough fuel for one pass. If we couldn't drop on that one pass, then we'd have to land somewhere. No Ranger wants to do that. I have far more landings when I should have jumped due to weather conditions—usually high winds that prevented us from dropping.

With my full load of ammo, PRC-77 radio (staff officers carried their own radios, no private to carry it for us), spare batteries, main and spare parachutes, and the other gear I needed, I was carrying more than double my body weight. When the jumpmaster gave the command to stand up, all the weight was centered under my reserve parachute, creating tremendous strain on my low back. There were many Rangers on this operation carrying far more weight than was I.

Ten minutes! We were jumping from C-141 Starlifters, cargo jet aircraft and about 100 Rangers in each bird. The doors opened, we shuffled forwards to get ready to exit and the jumpmaster called out, "Winds on the drop zone, 30 knots!"

Peacetime regulations limit airborne operations when winds exceed 13 knots due to the increased likelihood of injury from high winds.

The jumpmaster blew across his open palm, then flashed all ten fingers three times. Winds were thirty knots! Then he pointed skyward and circled his finger, indicating that we were doing a racetrack — we were going into a ten-minute circle so the aircraft could make another run at the drop zone. Every Ranger on those five birds groaned in unison.

Due to the quick (ten minutes with that weight on you is an eternity) racetrack, we had to stay standing, in position and ready to exit. As we approached the drop zone, the jumpmaster again blew across his palm and again flashed the signal for thirty knot winds. Another racetrack. An officer beside me raised his eyebrows, the unspoken question being, "How much fuel?"

This time as we neared the drop zone, the jumpmaster blew across his palm and held up ten fingers, once. We were a go.

As I exited the aircraft, it was dusk and visibility was good. I could see the Rangers in the lead aircraft were already hitting the ground, and their canopies were dragging them across the drop zone. We were jumping onto

Ali-Al-Salem airfield, which had been destroyed by our bombing campaign. The runway had been patched and was in use by allied forces, but most of the surrounding buildings were still in rubble, and debris littered the area. Years later I obtained a copy of a video taken by a ground unit watching us jump, and as the camera turned it was clear the wind sock extended straight out and the camera lens cover, secured to the camera with a cord, flapped in the wind.

As I descended, I could tell I was traveling faster laterally than I was falling, not necessarily a good thing. I saw a steel 55-gallon drum lying on its side, shot full of holes and crushed mostly flat. It looked like a jagged cheese grater. A T-10 parachute affords little to no control over where you go and I was headed right towards the drum. I dropped my ruck, planted my feet firmly atop the barrel and pushed off, rolling to the other side. Immediately my parachute caught the wind and I was drug about a hundred yards before I could release one of the suspension risers and collapse the chute. I was lucky. We jumped 450 Rangers that night, and over forty of them ended up in the hospital with concussions and broken bones. Many of the others sustained injuries they never reported because they didn't require evacuation and emergency care.

An all-night march through cleared paths through minefields, a resupply drop the next day where the winds were so high that water bottles littered the desert for a kilometer, then another all-night march. We blew up an objective, combining mortars, artillery, aircraft, and small arms fire in a coordinated orchestra of destruction. The next day we walked about twelve kilometers (ten miles) and boarded buses which then took us down the highway of death to Camp Doha in Kuwait City.

The highway of death was an awful expanse of highway strewn with vehicles, all headed north from Kuwait City and destroyed by allied aircraft. A quick Google search, "Highway of Death" and you'll get a small idea of what it was like, but unless you saw it, it is not comprehendible in its entirety.

I carried a historical heirloom with me on that operation. My Great-Uncle Roy had died earlier in the year, and among the items I inherited was the canteen cup he carried in combat during the Meuse-Argonne Offensive in France. It was the exact same shape and size as the canteen cup the Army had issued me, indistinguishable except it was steel rather than aluminum and for the date stamped into the bottom: 1918. I replaced the modern version with his, and I jumped with it into Kuwait.

As the battalion S1, it was my responsibility to find out what if anything was authorized the battalion for deploying into a combat zone, for that area was still officially part of Desert Storm, the campaign called, "The Liberation and Defense of Kuwait." I learned we had been in the theater of operations long enough to qualify for "hostile fire and imminent danger pay," $110 for the month, and anyone qualifying for hostile fire pay also qualified for the Southwest Asia Service Medal with a bronze campaign star.

Once we returned to Hunter Army Airfield, I issued a memorandum for the file of every Ranger who participated in the operation.

As personnel officer, I was authorized to sign documents "For the Commander". The memorandum stated, "I certify the following Rangers deployed in support of Operation Iris Gold." It included a by-name list, and concluded with the statement, "These Rangers are entitled to hazardous duty pay for the time period covered. Additionally, they are authorized the Southwest Asia Service Medal with Bronze Star and are eligible for wartime veterans' benefits."

Several years later a subsequent regimental commander issued orders awarding all Rangers who participated in the exercise the right shoulder combat patch, crediting each of us with a combat deployment.

While in Savannah, we had joined White Bluff United Methodist Church. White Bluff had a dynamic young assistant pastor, named Randy. Randy formed a group of men that met once a week at 4:00 a.m. His challenge to other men—"What are you doing then, sleeping?" I was normally at work, so it worked just fine for me. I invited another officer from the battalion and he joined me as we helped lead this small accountability group of men. We used a text written by Dr. Abraham A. Low titled, "Mental Health Through Will-Training" and the main principle I brought from this group and now use in my work with veterans to this day, is the concept of "response-ability": the ability to choose one's response to any situation facing them. I'll address this further in Chapter 11, Response-Ability.

Before our deployment to Kuwait, a promotable captain reported in to the battalion. Our new officer was Capt. Joe Votel, soon to be Maj. Votel. I personally guided him through the process of drawing his gear and getting his assignment paperwork in order, and I would later work for him. Joe went on to command the 1st Ranger Battalion and on 9/11 he was the Ranger regimental commander. He led the group of Rangers that parachuted into Afghanistan in October of 2001. When I made the fateful decision to leave the Army, he was the first one I told.

After nearly a year as the battalion personnel officer, in June of 1992, Lt. Col. Stauss called me in and told me I was being moved to the assistant operations officer (AS3) and that I would be working for now Maj. Joe Votel.

I was excited to be returning to the operational side of the Army, and especially in the 1st Ranger Battalion. Whereas the personnel officer was always worried about the suspense for awards and evaluations, the AS3 got to do the sexy stuff—the training. I had performed this job twice before, but at the brigade level—a higher staff level, when I was stationed in Hawaii.

Ken Stauss had already taught me time management. Now I would get to learn at his side while we evaluated all the combat platoons in the battalion. Over the better part of a month, we put each Ranger platoon through a demanding exercise involving moving a significant distance through dense forest, to find and destroy an enemy position.

Talk about a dream come true. I walked by the side of a Ranger battalion commander and helped him tactically evaluate every combat platoon in the battalion. Together we learned the strengths and weaknesses of every component of the unit. I benefited from his experience as we willingly shared his lessons learned, both with me individually and with the units we evaluated. Few officers in the Army get this opportunity, and I was one.

In October of 1992 selected leaders from the 3rd Ranger Battalion came to Savannah and matched up with their counterparts from the 1st Ranger Battalion. We were embarking on an operation called Embryo Stage. It was rumored to be the rehearsal for an operation overseas—a real-world operation, perhaps to snatch a bad guy, or to obtain evidence of weapons of mass destruction. This was reinforced by the fact that we had leaders from another battalion on the exercise with us.

In the air mission briefing a few hours before we launched, I sat across a conference room table from Lt. Col. (P) John Keneally. The "(P)" after an officer's rank indicated they had been selected for the next higher rank but the date to assume that rank had not yet arrived. Keneally was the commander of the 3rd Ranger Battalion. Next to him was US Air Force combat controller, Capt. Michael Nazionale. Also seated at the table were Ken Stauss and a dozen other staff officers.

We reviewed the details of the movement to the training area, Dugway Proving Grounds, Utah. The battalion would move by C-141 jets, with a portion going in advance, landing at Hill Air Force Base and changing to MH-60 Blackhawks, the Air Force special operations version of the Army's workhorse. The plan was for the assault force to hit a runway at Dugway, parachuting in if necessary or landing by helicopter if possible; then the rest of the unit would land at the newly secured airfield, launch a follow-on mission to secure the "precious cargo," then collapse all the forces back through the airfield and ultimately back to Hunter Army Airfield for a debrief and After Action Review. Elements of at least three branches of the service were participating.

This night was complicated by the most intense bad weather imaginable. I didn't have a full appreciation for how bad it was, for I was flying in a C-141 timed to arrive at the airfield several precise moments after it had been secured.

What followed was mentally, emotionally, and even physically stunning. I had been listening in on a satellite radio when the report came over that a helicopter had crashed into the lake. Immediately our aircraft turned and landed with the others at Hill AFB. I remember the ramp door dropping and monsoon-like rain falling outside. After some time, we took off and returned to Hunter Army Airfield in Savannah. We offloaded at a secure area of the airfield, one where Rangers plan operations away from prying eyes and ears. We were sequestered there until all the aircraft returned and we could positively identify who was missing.

In the chaos of executing the mission, Rangers and Airmen had been bumped from aircraft and some had clawed their way onto others. No one wants to be left behind.

It was nearly twenty-four hours before we confirmed the awful truth of who was on that bird. That bird carried two Ranger battalion commanders, an Air Force special operations squadron commander, a Ranger 1st sergeant, two communications experts and five Air Force special operations Combat Controllers. It is a rule written in stone, seemingly by the hand of the Great Ranger in the Sky Himself that key personnel will never, ever, be on the same aircraft or in the same vehicle to prevent exactly the type of disaster that wiped out three special operations senior commanders in one fell swoop.

These were men who were slightly senior or peers to McChrystal, Petraeus, McRaven, and Casey—all military leaders holding defining positions in the wars following 9/11. Their deaths left a gap in the ranks of the most gifted leaders in the military. Had they survived would our overseas wars have been different?

Killed on the bird that day:

From the US Army: Col. John Keneally, Commander, 3rd Ranger Battalion; Lt. Col. Ken Stauss, Commander 1st Ranger Battalion; 1st Sgt. Harvey Moore, C Company, 1st Ranger Battalion; Sgt. Blaine Mishak, 1st Ranger Battalion Commander's RTO; and Spc. Jeremy Bird, 1st Ranger Battalion Commander's RTO.

From the US Air Force: Lt. Col. Roland Peixotto, Jr.; Capt. Michael Nazionale; Tech. Sgt. Mark Scholl; Staff Sgt. Steven Kelley; Sgt. Philip Kesler; Sgt. Mark Lee and Senior Airman Derek Hughes.

Every single soldier present that fateful day could have been any one of them. Every one of us who survived often wonders why and whether the world would have been better off if our places were switched.

Not a day goes by that I don't think about and remember Ken Stauss. I wear a tribute bracelet with his name on it every day. When I fly someplace the last thing I do before turning off my phone on the plane is to text Sara, tell her I've boarded and that I love her. Only after working with veterans for a few years did I realize why I did that. Of my sixteen friends that died on active duty, all but one had died in aircraft incidences. Should that happen I wanted my last message to her to be one of love. We don't always realize how our experiences continue to shape and effect our thoughts, emotions and actions today.

One of the Rangers on the mission, Wesley Jurena, wrote a blog post years afterward that captures some of the rawness of the night. He was originally on the aircraft that crashed but was bumped to another aircraft shortly before the mission began.

Access this blog at www.KarlPMonger.com.

Years later a movie came out about a female officer, a helicopter pilot who is shot down in Iraq with a bird full of soldiers. I had to leave the theater during the crash sequence.

Lt. Col. (P) Keneally would be posthumously promoted to full Colonel.

Twenty years later, GallantFew sponsored a commemoration at the crash site, just off Antelope Island in Utah's Great Salt Lake. There is a Ranger monument there, containing pictures of the twelve killed, a monument envisioned and driven by the passion of the father of one of the Rangers killed.

Many of the families came. It was the first time many of them had been to the crash site, the first time many of them had met each other. At a reception the night before the ceremony at the monument, we gathered in a room at a local hotel. I was pleasantly surprised to see a few comrades of those killed who came, and a group of Air Force Special Operators stood in a circle, drinking beer and telling stories. Then a young woman walked through the door.

She stopped, looked around, and said, "Did anyone here know my dad?"

She was six years old when her father died. She had come alone, to try and touch a tangible memory of her father.

One of the men said, "He was my best friend."

This is still so raw to me that I have tears in my eyes while I'm typing.

We all cried.

She produced a picture of her father, a young man smiling happily for the camera, and she asked what the story was behind this picture because he looked so happy.

The men shifted around, slightly uneasily. Then one said, "We were at a night club. Let me tell you about your dad."

Tears turned into laughter, and the stories flowed.

Chapter 7

Back to Work

Back to work.

Death is not an unusual event in the military, one which occurs all too often. A common saying goes, "A veteran is someone who has written a blank check to the country – up to and including his or her life."

That's reality. Whether a young man or woman volunteers to serve in an elite special operations unit as a door-kicker or becomes a mechanic in a maintenance unit – in any branch of the service a volunteer incurs the risk of being sent into harm's way. Danger in training, danger in a warzone. Besides the twelve that died that night, another four men with whom I served gave their lives wearing the uniform.

There isn't a lot of time to mourn in the military. We still had a job to do, we were the point of the spear – the force that when the President called, we had to be wheels up, on a bird flying to an emergency anywhere in the world within eighteen hours. Even so, for the next several weeks we operated in a fog. The loss was so terrible, the grief so profound.

The Regiment held memorial services, and the staff officers flew to Fort Smith, Arkansas where Ken was laid to rest just six feet from Brig. Gen. Darby—the officer who in World War II had the vision to start the Rangers. The slow pace of the caisson, the sound of the bagpipes, are forever seared in my memory and to this day I cannot hear Taps without my eyes growing hot and filling with tears.

As we reeled from the loss, the Army looked to rapidly fill the gap in Ranger Regiment leadership. Two Ranger battalion commanders were gone in an instant. In the space of one year from the crash, the 3rd Ranger Battalion would fight in the streets of Somalia.

Ranger battalion commanders must first command a battalion in the regular Army before commanding in the Regiment. We got an interim battalion commander, but because he had not commanded another battalion he was not allowed to stay long term (he went on to command the 2nd Ranger Battalion and ultimately retired as a Lt. Gen.). During the four months he was there, we continued to focus on combat readiness. The pace doesn't slow in a Ranger battalion.

After a few months, we learned who our permanent battalion commander would be, and it was an officer I'd served with before, who had heard me ask to not be left behind were the battalion deployed to Desert Storm.

By now I had two officer efficiency reports written by two Ranger battalion commanders (the interim had been there long enough to rate me) and both had written in my evaluation I would command a Ranger company. The most fulfilling job of my military career so far had been commanding B Company, 5th Battalion, 14th Infantry and I was eager for the responsibility of command again.

A few weeks before I was to assume command, the new commander called me in his office and told me that he was assigning command of the company to another officer. I was devastated. Jana was equally upset; she was eager to move back into the role of family support group leader, one at which she excelled. It was the first time the thought to leave the Army had crossed my mind since commissioning. With two mentions of Ranger company command in my record, then no command, I realized that I might never command an infantry battalion.

As for that lucky Ranger officer who took that company command, at the writing of this book he remains on active duty as a Major General. He is the epitome of a professional and has my utmost respect.

A briefing book circulates among the primary staff officers every morning. Any messages that arrive overnight are classified by staff section, personnel, intelligence, operations/training, etc., and after the battalion commander and executive officer have reviewed it, it is passed from staff section to staff section. As assistant S3, I was high in the queue to read messages, and the morning after I lost my dream job I turned the pages of the binder, still unsure of my fate. There in the personnel tab was a message from the Department of the Army. Not enough captains in my year group had volunteered for the Voluntary Separation Incentive Program (VSI).

Officers in the Army are identified by their year of commission. Since I was commissioned in 1983, I was Year Group 83 (YG83). An officer remained in that year group his or her entire career, unless picked up "below the zone" and promoted a year earlier than contemporaries, effectively changing his or her year group. Not enough captains in YG83 had volunteered for separation, and under President Clinton's reduction of the military, many of us had to go. The Army first asks officers to leave voluntarily, and pays a decent "bonus" to leave. The idea is to get enough volunteers so that the Army doesn't have to force officers out. I did the math, and to me it was a sign from God.

When I was an ROTC cadet, one of the cadre members, an active duty Army major, sat down with me and gave me career advice. He told me, "If you want to be a success in the Army, if you want to make General, stay middle of the road. Don't be a paratrooper, don't be a Ranger, don't go special forces. All that stuff is fancy and dead-end." This seemed to confirm to me that I'd derailed my military career.

Boy, was he wrong.

I violated one of the cardinal rules that I have since learned. I did not consult any of my mentors. My pride was so hurt, my self-confidence so shaken, that I couldn't bear the thought of contacting some of the officers that had put me here, to ask their opinion on whether I had a viable future in the Army.

Look at the most senior Generals in today's Army. A far disproportionate number of them have service in the Ranger Regiment.

In 1974 the Army was reeling from defeat in Vietnam (historians tell us we had won the battles but not the war), and the Army's discipline was in shambles. Drug abuse was rampant and soldiers disrespected their leaders. The Army was trying to transition to an all-volunteer force as the draft ended in 1973. The Army was a hollow force.

Then Chief of Staff of the Army General Creighton Abrams, who earned his place in history playing a role rescuing American forces trapped during the WWII Battle of the Bulge, issued a prescient document known as "Abrams' Charter."

Abrams' Charter was a visionary document intended to rebuild the damage done to the Army during the Vietnam years.

http://www.dtic.mil/cgi-bin/GetTRDoc?AD=ADA415822

Abrams' Charter
General Creighton Abrams, Jr.
26th Chief of Staff
United States Army

"The Ranger Battalion is to be an elite, light and most proficient infantry battalion in the world. A Battalion that can do things with its hands and weapons better than anyone. The Battalion will contain no 'hoodlums or brigands' and if the battalion is formed from such persons, it will be disbanded. Wherever the Battalion goes, it must be apparent that it is the best."

In a subsequent Chief of Staff Charter, Gen. Wickham directed that soldiers assigned to this unit would not be allowed to "homestead" or stay there for their entire career, rather they must rotate back to the regular Army periodically. In so doing, he began the process of seeding the Army with soldiers who demanded excellence, who knew how to train, who were extremely physically fit and who knew how to develop their subordinates.

Wickham's Charter
General John A. Wickham
30th Chief of Staff
United States Army

"The Ranger Regiment will draw its members from the entire Army. After service in the Regiment, return these men to line units of the Army with the Ranger philosophy and standards. Rangers will lead the way in developing tactics, training techniques, and doctrine for the Army's light infantry formations. The Ranger Regiment will be deeply involved in the development of Ranger doctrine. The Regiment will experiment with new equipment to include off-the-shelf items and share the results with the light infantry community."

I can testify to the reality of the Army's condition and to the effectiveness of Abrams' vision. A few years before I arrived at my first assignment in the Army, my battalion experienced this first hand. A soldier under disciplinary action attempted to murder the battalion commander by shooting him while he was riding a bicycle. It was still being talked about when I arrived there. My first company commander taught me how and where to look for and identify illicit drug use. My first day as a platoon leader our company had a surprise inspection and we found stashes of contraband — drugs, weapons, even stolen military gear.

On the other side of the post airfield, however, was a unique unit. The 2nd Ranger Battalion, famed for scaling the cliffs of Pointe du Hoc during the Normandy D-Day invasion, had been reactivated and stationed there in 1974. These soldiers were distinctive in appearance as they wore starched olive drab green jungle fatigues and black berets, the only soldiers in the Army authorized to do so. They had just jumped into the pages of history during the invasion of Grenada, where they jumped so low that they removed their reserve parachutes, as there wouldn't have been time to employ them if needed and it was extra bulk to carry.

Occasionally we received one of these soldiers who had been injured, or who had been released by the unit for one reason or another, and they were in demand — because they were great soldiers. Soon soldiers from my platoon were asking permission to volunteer to apply to go to the Rangers.

Today's senior leadership, from general officers to sergeants major contains representation from the Ranger Regiment in vast disproportion to the Regiment's size compared with the overall Army.

Each subsequent Chief of Staff of the Army has issued his own version of the Charter, reinforcing the importance of the Ranger Regiment and adding additional guidance relative to the Army and current threat conditions.

Even the prevalent Family Resource Group, or FRG, began in the Ranger battalion and now is the model for providing family support for every unit in the Army.

I believe the overwhelming success of the military in Desert Storm was a direct result of the impact Abrams Charter had in rebuilding the Army. One last note on this: when I first entered the Army, only Rangers said "Hooah." Now a soldier from any component of the Army goes on TV, what do they shout? "Hooah."

The very best leadership and training techniques I learned in the Army came from officers and non-commissioned officers with Ranger experience. I brought all these to bear as a company commander, and twenty-two years later I received a surprise phone call from an active duty command sergeant major. Nicknamed "short-round" by his peers many years before when he was a private in my command, Troy was now the senior enlisted advisor taking a battalion from the 101st Airborne for a combat rotation in Iraq.

Troy called to tell me that the training techniques and standards that I and Jimmy Akuna had instilled back then were the very techniques and standards that he used today, and he wanted to know that his soldiers were well trained, and he gave me and my 1st Sergeant credit.

He took his battalion to combat and brought every one of them home a year later. There is no higher professional compliment that I would ever receive, but I must pass the credit along to the Rangers who trained me: Hale, Dubik, Akuna, Ohle, and others.

Rangers are universally respected within the military and when one shows up, other soldiers look to him for leadership. It's no different on the veteran side of service. When one of these highly respected warriors stands up and shares a story of how they overcame Post Traumatic Stress (PTS), or how they deal with a Traumatic Brain Injury (TBI), or many other transition After Action Review (AAR) points, others will take notice. "If a Ranger can reach out for help, so can I."

This is a call for Rangers to share their successes and their struggles and to be leaders in their communities. Even Brig. Gen. Darby urged this in his farewell letter.

"To the Officers and Men Who Served with
the First, Third and Fourth Ranger Battalions"
after they were disbanded in 1944:

When this war comes to an end, most of you will return to the way of life which you fought so hard to return to — to pick up the threads of your civilian pursuits. You will bring back with you many nostalgic memories of your fighting days — both bitter and pleasant. But above all, you will bring back with you many personal characteristics enriched by your experiences with the Rangers. In whatever field or profession you may follow, I know that you will continue as civilians with the same spirit and qualities you demonstrated as a Ranger. Your aggressiveness and initiative will be tempered to adjust to civilian life with little difficulty. In your hearts as in mine, you will always have that feeling ... of being a Ranger always."

Abrams' Charter did exactly what he intended and far beyond what he could have ever imagined. Of the twenty or so captains I served with there, more than half of them have become General Officers. In the other battalions in which I served, the count is still none.

Again, my destiny would lie elsewhere.

I did the math. There were two options: a lump sum, one-time payment of about $40,000, or an annual payment calculated at a percent of base pay times twice the number of years served. For me, it was $9,700 a year for twenty years. There was little time to waste; the window was open for only two weeks to apply. I submitted my packet.

A couple days later I received a call from an officer from the Pentagon's personnel division. He tried to talk me out of leaving, telling me that I had a bad experience with one battalion commander but that wouldn't be a career ender, but I wasn't interested in anything he had to say. Within a few weeks I was out-processing from the Army. There was no transition assistance program to speak of, and I grabbed a copy of *What Color is Your Parachute* and started planning my move to the next stage of my life.

Soon I received a call from a well-known military junior officer placement service, an organization known as a "headhunter." The person I spoke with was himself a former Army officer, and he noted my Ranger staff officer experience and my Top Secret/Special Compartmented Information security clearance. He told me they would find me a great, well-paying job and explained how it worked.

They would arrange the job interview, I would fly there and if they liked me, I would agree to take the job.

I said, "Great! As long as it is within a half-day drive of Wichita, Kansas." After ten years as an Army officer, three of which had been in a Ranger battalion and subject to deployment at any moment with no notice, I wanted my kids, now ages eight and six, to get to know their great-grandparents. Both my maternal grandparents were healthy, and along with Jana's, were living in Wichita. I wanted my kids to know these grandparents and to develop a sense of family they had never known. The recruiter told me that wasn't how it worked, that I'd have to go wherever they found the position, and I told him that I wasn't interested in working with them. I headed to Wichita with no retirement, no income other than the annual stipend of almost $10k, the first payment due in a few months, no health insurance, and nothing but uncertainty.

On one of the last days at battalion, I was called into a staff officers' meeting. The battalion commander presented me with my second Meritorious Service Medal, and the personnel officer (Steve Banach) read a summary of my accomplishments at the unit. In the three years I was there I had overseen the revisions of the Readiness Standard Operating Procedures (SOP) — the document that laid out every detail needed for the battalion to deploy; the Personnel SOP; the Garrison SOP (everything needed to maintain peacetime operations at home); the Airborne SOP (governing all parachute operations); and the Tactical SOP (how the battalion fought). He pinned the medal to my uniform pocket, stepped back, looked me in the eye and said, "You really did all those things?"

"Yes, I did."

I think I knew more about how that unit functioned than most anyone else there, and now I was going to try and forget all that and just go do something else.

I felt like a complete failure. I'd lost the opportunity to command a Ranger company, I'd given up my military career halfway to retirement, and was too ashamed to reach out to those who knew me best to ask for their guidance and help. I didn't want to go through the shame of explaining why I was leaving the Army. It took several years for me to understand that these feelings are not uncommon, especially among soldiers that serve in elite units and leave for any number of reasons.

The Regiment has a Blue Book, it details the Ranger Standards. This small blue pamphlet lays out everything expected of a Ranger, and while it's probably changed since my day there, I still have my copy. Failure to meet any of the standards in the booklet can result in being RFS'd, Released for Standards. A Ranger can be RFS'd for operating a vehicle under the influence of alcohol, for having an accidental discharge (Rangers carry their firearms locked and loaded, ready for action. Even in training, accidentally discharging a round—even a training blank—is grounds for automatic dismissal). While I hadn't been RFS'd, I sure felt that way.

Today, I encounter soldiers who have left active duty under a wide range of circumstances, many of them less than ideal. Improvised Explosive Devices (IEDs), explosive door breaches, even the concussive effects of our powerful weapons, such as the Gustav recoilless rifle (which has an overpressure so great that in peacetime soldiers are not allowed to fire more than one round a day to avoid damage to internal organs) create blast waves that rock through brain tissue causing traumatic brain injury (TBI). Virtually every soldier experiences TBIs—from mild during training to severe during IED attacks. TBI symptoms include short term memory loss, mental confusion, difficulty speaking and comprehending, and more. These conditions have resulted in soldiers misbehaving, underperforming, and behaving in uncharacteristic ways.

Soldiers who misbehave get punished. They don't get promoted, they don't get to re-enlist. Some of them are summarily dismissed from the military. If the misbehavior is great enough—illicit drug use, or abuse of prescription medications—they might earn an Other-Than-Honorable discharge. Most employers now ask on their job applications whether a veteran has an honorable discharge. If a veteran presents a General or Dishonorable discharge, it's tough to get hired. If the discharge falls to the worst category of dishonorable, the veteran loses education and VA benefits as well.

Some of the veterans who need help the most are denied it by the very organization charged with providing it. There is late breaking news on this topic, as while I write this today (2017) there is an announcement that the new Secretary of the VA intends to open mental health care to veterans with other-than-honorable discharges or in crisis. It probably astounds you to know that the VA would not already offer mental health care benefits to *any* veteran in crisis.
    https://www.va.gov/opa/pressrel/pressrelease.cfm?id=2867

It gets worse.

Soldiers who are injured or wounded to the point where they will be discharged, are assigned to Warrior Transition Units (WTUs). There are absolute horror stories tied to WTUs. The typical WTU is not staffed by our best and brightest soldiers. Those soldiers hold positions of command in combat units. WTUs are staffed by reservists, many of whom are activated for this purpose and have little to no combat experience, and little time in a field unit leading troops. Perhaps you see where this is going.

A combat hardened young PFC assigned to a WTU pending medical discharge or retirement is told to sweep floors by a sergeant without combat time. The coming conflict is predictable and preventable, except it isn't. While the authority of rank is assigned by the Army, respect is earned. The Army likes to talk about a program called "Soldier for Life" in which it encourages soldiers to remain affiliated with and proud of the Army for life. Poor leaders who abuse their positions and refuse to treat soldiers with respect, especially those wounded or injured and going through the medical separation process, have the exact opposite effect of making a soldier want to affiliate with the Army "for life." I've worked with soldiers who suffered broken backs but were required to stand at the Charge of Quarters (CQ—the admin entry point to a barracks) desk the entire night because the other soldiers needed to train during the day, and he wasn't doing anything. Never mind the incredible pain and distress he was put through, and if he complained he was ordered to do pushups. There is no excuse for toxic leadership like this.

It's not just privates who receive this deplorable treatment. I know a senior non-commissioned officer who suffered terrible injuries, and takes a potent cocktail of pain, sleep, and anxiety medications. The drugs knock him out for a good eight hours, and he needs the sleep for his body to heal. He can't drive due to damage to his eyes and arms, and when he was assigned to a WTU he was incapable of dressing himself. His caregiver mother provided round-the-clock care for him, and due to his status and rank, he owned his own home away from the WTU barracks.

The sergeant in charge of his unit ordered him to attend daily 0600 (6:00 am) accountability formation. To be there by 0600, my friend (and his mom) had to wake two hours prior to formation time, get dressed, and drive to the WTU. The meds knocked him out for eight hours, so he had to be asleep by 2000 (8:00 pm). Going through his ritual of getting ready for bed, dressing wounds and taking meds took another hour or two. He had to start getting ready for bed at 1800 (6:00 pm). Not only him—but also his mom.

My friend told his sergeant that he would not attend the daily formation, but he would call in so the sergeant could satisfy accountability. The sergeant threatened my friend with military punishment. My friend told him to pound sand—and he ended up prevailing because of his rank and experience.

What's a private to do?

What kind of leadership allows this to happen?

If we valued those who have paid the highest price short of giving their lives, we would assign the best and brightest in our military leaders to care for them. An officer selected to serve in the Ranger Regiment should serve three or six months in a WTU. Instead, we put the least qualified, least experienced in those roles.

I have too many stories from too many friends about these abuses, and in 2015 a major news article was published about these abuses in Texas and Kansas.

http://www.nbcdfw.com/investigations/New-Records-Show-Injured-Soldiers-Describe-Mistreatment-Nationwide-From-Commanders-at-Armys-Warrior-Transition-Units-WTUs-298816361.html

The military wasn't prepared for the number or severity of these injuries, and the soldiers weren't trained in how to identify them, report them, and deal with them. How many since October 2011, when we invaded Afghanistan, have suffered injuries like these and subsequent mistreatment in the system?

While my circumstances departing active duty were relatively on my own terms and with a medal pinned to my chest, I still experienced shame and regret at the way my service ended. How much more so for these young men and women who have endured TBI and PTS, and were classified as dirt bags and sent packing?

Three years in a Ranger battalion can wreak havoc on a family. We were subject to deployment at a moment's notice, and that meant anytime, to anywhere. Many units practice their deployment notification sequences in the early morning hours, to catch people by surprise and ensure the phone call trees worked (no cell phones then and pagers were just becoming widespread). The reality is that a Ranger could be alerted and deployed at 5:00 p.m. in the afternoon—and just not come home from work. When an alert sequence started, all non-secure communications to the outside world would be severed. There were no phone calls home to say I'm working late, no warnings.

Jana and I had developed a system. If I didn't come home from work, the next day she was to drive to the battalion area and look for my car. If there was a note in it saying I'd gone for chocolate, that meant I was gone and would be home when I came home. I only had to leave that note once.

I'd put my family through so much, now I owed them stability. I owed them the opportunity to know their family, and I also knew the folks in Wichita desperately wanted to grow close to my kids. I had such a close relationship with my grandfather, and I wanted that for my kids too.

The recruiter's rejection was a hard lesson. Business opportunity came first, family second. Sorry, 'been there done that. He cared less what I wanted than about satisfying his client and getting paid. I understand this now having been a commissioned salesperson myself. I might have passed up the job of a lifetime, but I'd had that already and I was slipping into the dangerous mindset that I would never be part of anything so amazing professionally again—and I was wrong. This shaped my approach to help veterans today. I want to know where *they* want to live, what is important to them, and I want to help them discover there is purpose after military service.

Chapter 8

Now What?

Back in Wichita, my head was spinning. The four of us had moved into my parent's basement while I searched for a job. One of the first interviews I landed was with one of the largest privately held companies in the U.S., headquartered in Wichita. A human resources manager looked at me across a table, glanced at my resume and said, "Hmm, Army officer. You guys are really good at taking orders and doing what you are told, but here at this company we need people who have initiative, think outside of the box and don't need constant supervision."

Wow.

I packed up my materials and left, thinking to myself that I could do her job blindfolded, with one arm tied behind my back. How could she possible not recognize the raw talent sitting in front of her? Didn't she know what the former personnel officer and assistant operations officer for a Ranger battalion could accomplish?

She didn't. The human resources world hadn't prepared her, and the military hadn't prepared me to help her understand.

It's not the military's job to make soldiers into civilians; its job is to take civilians and make them into soldiers, and to use those soldiers to force other nations to behave the way we want them to, either by the use or the threat of force. That's why commanders shed soldiers that don't make the unit more lethal, just as I had done as a company commander in Hawaii.

Communities must take responsibility for bringing soldiers home and helping them become contributing community members. Veterans who are established in the community must step up and take leadership roles in making this happen. Our country has invested tremendous resources in providing military men and women proven, results-oriented leadership and management training and has given them responsibilities and authority far beyond that found in the civilian workplace especially at a young age. These skills are perishable, and the longer that veteran's transition process continues, the more discouraged the veteran becomes and the possibility of isolation grows larger. I'll address this further in Chapters 14 and 15.

I thought I knew transition. I had read *What Color is My Parachute*, fine-tuned my resume and, following the guidance in the book, started networking.

Soon I had an interview with a family owned business, a soft drink bottler. I got the interview because my brother-in-law's father was friends with the owner (see that networking was paying off!). It turned out that the owner of the company was a Korean War-era infantryman so we hit it off right away. The company didn't have any open positions, but after talking with me for a few moments, he asked, "What would it take to hire you?"

I had done my homework. I know (remember, this is 1993) that to cover our projected household expenses, and considering the $9,700 I would receive annually from the Army, $25,000 would be just barely enough. I said, "$30,000" and he said answered, "Done."

Shoot, I should have asked for $40k.

This company ended up creating a position for me, Safety and Training Director. In my experience, the best jobs aren't found by looking in the want ads, or online job boards, they are found by networking. Many times the position is never formally advertised, and is filled before many people even know the opportunity exists. This was a great opportunity for me, and over the next two years I was promoted several times, ultimately with responsibility for distribution of beverages over a two-state area.

But I found myself having a hard time fitting in. I left an environment where as one of the senior Captains in a Ranger battalion, when I entered a room everyone stopped what they were doing and stood up. Here as safety director I zealously went about my duties, ensuring our employees abided by laws and regulations. I'm telling workers in the pressurization room to put on their safety glasses, I told a senior leader in the company he couldn't smoke in the production room. The responses I got were significantly different than what I would have received in the Army. I was told where I could put the safety glasses, and I was ignored by many. I even picked up the nickname, "Deputy Dog."

One day while walking past the vehicle mechanic bay, I noticed a new mechanic. I stopped to talk with him and soon learned that he was a Desert Storm veteran, "Derrick" (not his real name). Interaction with Derrick was my first experience with a young veteran suffering post-traumatic stress (PTS). One of my great-uncles, a World War II veteran clearly had it but I didn't understand it at the time.

I have learned and believe that post-traumatic stress is a natural reaction to an unnatural situation. It is not a disorder. It's an injury, just as a bullet tears flesh, but it leaves no mark. Because it's an injury, however, with the right treatment, it has the potential to heal. Unfortunately, the generally accepted method the VA uses to treat PTS, called "immersion therapy," is akin to ripping the scab off the wound to see if it's healed. All but a few of the veterans I know who have tried immersion therapy drop out after a few sessions, because it's too painful. I heard about one situation where a veteran experienced severe PTS whenever he heard loud noises, so the VA counselor told his wife to set up firecrackers outside the house at random times until he got used to it. More on this in Functional Emotional Fitness, Chapter 13.

Derrick confided in me that there were people watching him, hunting him, and that he had to be extremely vigilant so that he wouldn't be attacked. He was jumpy, and he was self-medicating. I made a point of visiting with him every chance I could, and tried to understand what he was going through.

We developed a friendship, and he struggled to get the VA to recognize his PTS. After about a year he called late one night. He was extremely agitated and upset, and wanted to come and stay at my house. I knew he needed more help than I was prepared or able to give him, so I refused. I ordered him to go to a treatment facility and ask to be admitted, and he did — and they did. I am extremely proud of Derrick, and later hired him to work as a salesman for me at another company. He went into the National Guard and became an officer, and after years of struggling, was justified when the VA finally recognized his PTS as a total 100% disability. He's now pursuing his dream of becoming a chef and opening his own restaurant someday.

To keep one foot in the military and maintain my connection to standards and respect, I joined the reserves. When I walked in to the personnel officer in my Class B uniform (olive green slacks with a black stripe, black shoes shined so one could see their reflection, and a short sleeve light green shirt with all my awards and decorations), I created a bit of a stir. I overheard one clerk tell another that I looked like a recruiting poster. Seems the Army Reserves wasn't accustomed to having an airborne Ranger officer walk in their door.

I stayed in the reserves for a few years, and I was tickled that I got to wear the "rolling W" patch of the 89th Army Reserve Command. It was the same patch my great-uncle had worn on his uniform as a doughboy fighting in France. I got promoted to major. I also learned about the disincentives the reserve system has that cause good soldiers to quit.

When I was on active duty, I was plagued by back pain. First as a company commander in Hawaii, during our big deployment exercise in Arkansas I had picked up a "casualty," a Staff Sgt. a half foot taller and weighing fifty pounds more than me. I threw him over my shoulder in a fireman's carry and started towards the evacuation vehicles, and I stepped in a small depression hidden by tall weeds. I felt something in my low back give, but I kept going. It seemed to get better and I took some Motrin. A few months later, back in Hawaii, I was in my office following physical fitness training. I bent down to untie my running shoe and my back locked up with the most intense pain I had ever felt to that point. I was frozen, couldn't stand up, and couldn't drop to my knees.

The same Staff Sgt. was sitting in a platoon office across the hall from my open door. He saw me stuck there and laughed and laughed. I called him a few choice names, which only made him laugh more. Finally, I was able to get to the ground and roll to my side until the spasm stopped. Ever have a charley horse in your calf? Imagine having three across your low back simultaneously. I slowly made my way to the medics; they gave me stronger Motrin and sent me to get checked at the Troop Medical Clinic (TMC). At the TMC they gave me stronger muscle relaxers and told me to go home and rest.

"I'm an infantry company commander; that's not an option."

I did have them schedule me for physical therapy, and a couple times a week usually over lunch or during late afternoons I would lie on a traction table, a belt around my hips and another my chest and the machine would try to pull me apart. I saw an inversion table and asked about it, they showed me how to use it and I liked it so much I bought a set of gravity boots—cushioned collars with steel hooks that lock around your ankles. I used them to hang upside down from the kid's jungle gym in the shared yard of our on-post quarters.

Hard charging soldiers in the military don't report injuries, they don't ask for time off. Instead, they treat it as best they can and they hide it so that they don't appear weak. Unfortunately, doing this doesn't get the injuries documented in medical records, so proving the injury to the VA later is difficult if not impossible. It also creates conditions where the injury worsens over time, without the proper care and treatment.

Sometimes the soldier self-medicates, and this can be abuse of over-the-counter meds, prescription meds, alcohol, or illegal substances. I was fortunate in that the visit to the

TMC that day was recorded in my records, and the VA found it years later when at the urging of a friend I used a VFW Service Officer to file a disability claim. Because this was documented, when I had a laminectomy (surgery to remove spinal material that is impinging on a nerve) and I was in the VA system, the VA placed me on temporary 100% disability for the month during which I recovered from the surgery.

The VA also awarded me a 20% disability rating for three service-connected conditions: 10% for low back pain, 10% for asthma, and 0% for a fractured ankle. Zero percent sounds strange, but that means the veteran receives no compensation for that condition, but the VA acknowledges responsibility for care of that condition, and if it worsens there is a process to increase the rating over time. At that time, a 20% rating meant I would receive a tax-free check for $166.00 a month.

Remember the separation pay that I was counting on to help with my income? The second year I was out of the military, I checked the amount to be automatically deposited in my checking account. It was a quarter of the amount I expected! I contacted the toll-free number given me with my separation documents and learned there are intricate rules governing military retirement and separation pay. I could not collect separation pay, disability pay, and reserve active duty pay (the two-week annual training pay was considered same as active duty). The pay system cross matched my income, and the VA stopped paying my disability because of the active duty pay — which to me is totally reasonable; if I'm able enough to go on active duty, I shouldn't qualify for disability pay. It would have been nice to know for planning purposes. Negative financial surprises are never enjoyable.

Then the pay system matched my reserve pay to the separation pay, and reduced my separation pay by the amount I'd been paid for the active duty reserve time. It became apparent to me that if I continued in the reserves I would in effect be doing that for free—and by then I had moved to another job and was being paid commission. I could make more money in one sale than I'd get for a weekend drill, and I'd be doing the two-week annual training and sacrificing the separation pay in the process. I've been asked why didn't I stay in the reserves for retirement, and the biggest reason is that I wouldn't be able to draw a retirement check until the system accrued the amount of the separation I'd been paid, and reserve retirement pay doesn't start until age 62.

I'd be playing Army for free, and I'd never see a retirement check. I hung up the uniform for good. It's unfortunate that the system creates disincentives like that for soldiers to continue serving. I've talked with others who had similar experiences and left because it didn't make financial sense to continue.

Back at the beverage company, my frustration increased. The company was undergoing ownership transition and while I enjoyed the friends that I'd made and the challenges I'd overcome, the uncertainty of the company's direction bothered me.

Around this time, my buddy, Bill Cooper, who I first met when we were in the eighth grade, started talking to me about working for him. I joke that the first time I saw Bill, I saw a big kid holding a bully in a headlock under the piano in the choir room and I told myself, "He needs to be my friend!"

Over the years, through high school and college, we became practically inseparable. We were both in ROTC at Wichita State and were commissioned as 2nd Lts the same day. Bill had transferred to Wichita State from a community college, and not all his credits transferred with him, so he stayed in Wichita another year before he went on active duty. I ruined the spit shine on his boots while I was in Ranger School: I sent him a letter from the desert phase full of sand. He opened it while in uniform and got desert dust all over his boots.

Bill knew his transition plan. He left active duty after commanding a mechanized company of Bradley Fighting Vehicles in Germany and he targeted the construction equipment industry. He first became the service manager of a construction equipment dealer then became a sales representative, and now he was being promoted to be the branch manager. He took me to lunch and asked me to consider taking his place as salesman.

I told Bill that sounded like a recipe for disaster! I knew nothing about the equipment, didn't have the highest opinion of sales people, and my only experience selling anything had been as a young boy selling candy bars to raise money for my Little League baseball team.

I'll never forget what he told me. He said, "I can teach you about the iron, and how to sell. What I can't teach are discipline, integrity, and the ability to learn. You've demonstrated all of those in the Army. If you decide this is what you want to do, you'll be fine."

At that time, he was a better salesman than I, so I went to work for him.

Although I initially resisted sales, I found I enjoyed it. The people one calls on in the construction business are for the most part extremely hard-working, salt-of-the-earth people. They shook hands and kept their word, and I made some great friends. Bill was right; he taught me the iron, he taught me sales, and he spent a lot of time helping me smooth out my rough edges.

Seems I still tended to talk to people like I was a Ranger captain and I outranked them. It wasn't something I did on purpose, and I didn't realize I did it. Even my mom mentioned that sometimes I spoke to my kids like I was giving orders. Once Bill pulled me into his office and gave me some strong mentoring on how to interact with some of the folks in the office who weren't moving as fast as I wanted them to move, and I listened and worked on it.

One of my biggest struggles was making money. Not in getting paid, but in maximizing my profit margins on a sale. If I knew the "standard" was a certain percent, I was used to meeting standards. Adding in a point or two (or more) when the machine had been depreciated through rentals or other means made me feel like I was unfairly lining my pockets. Bill worked extensively with me to help me understand market value, the functions and features of the machinery, and how to sell. At the end of the first year I had sold a million dollars' worth of equipment and was recognized for having the highest average profit margins in the sales force.

Over the next fifteen years I worked for Bill twice, and quit on him twice to go work elsewhere. We've remained best friends and Bill would later be instrumental in the creation of GallantFew, giving me the nudge I needed to get going.

The second time I worked for Bill, he had moved to United Rentals where he was a District Manager. He hired me to run one of the largest United Rentals branches in the U.S., and while I was on a job site with a salesman we heard the news report on NPR that an aircraft had struck a tower at the World Trade Center in New York. I called Bill, and met him at his house.

Together we watched in horror as the two towers collapsed and we knew we were now a nation at war.

A month later, I watched with pride and jealousy as elements of the Ranger Regiment parachuted into Afghanistan. I soon learned that two of my friends had led the way. Lt. Col. Steve Banach commanded the 3rd Ranger Battalion and Col. Joe Votel commanded the 75th Ranger Regiment. I was in the inactive reserves by now; post-surgery the VA had increased my disability rating and while I thought about trying to go back on active duty, I looked at my young daughters and knew that I would have to serve in other ways this time. I shot Joe an email congratulating him and telling him how proud I was of him and the boys. To my great surprise, he responded the next day. When he changed command in 2003, I received an invite to attend.

As a United Rentals branch manager, I had responsibility for over 60 employees. I was determined to hire veterans, and began looking for veterans to hire. Derrick, the mechanic from the beverage company was one of those, but finding others was virtually impossible. I called the local VA Regional Hospital, I met with the local workforce alliance, trying to find veterans I could hire but I couldn't get them to send any to me. I even called up the newly established Army Wounded Warrior (AW2) program at Fort Riley, Kansas (not the nonprofit), asking what could I do to help veterans transitioning back to the Wichita, Kansas area. I told them I knew lots of business owners, faith leaders, and nonprofit executives, and was confident I could help any veteran returning to Wichita with home modifications (heck, I ran an equipment rental store!), jobs, networking; you name it and I'd try and find a solution for it.

A little different than merely following orders and doing what I was told.

I couldn't get anyone to return my calls. When I did get someone on the phone, they were in no position of authority to make anything happen. I got the impression I was interfering with their day, with my offers to help.

Then someone told me that the local Big Brothers Big Sisters agency, the same agency I had been a kid in thirty years ago, was searching for an Executive Director. I checked it out and felt immediately called to throw my hat in the ring. I made a phone call, spoke with the current Executive Director who was moving to a newly formed umbrella agency over a larger area of Kansas, and he invited me to come and talk with him. Next thing I knew, I was in front of several board members, and I had a job offer.

I went to Bill's office, dreading the thought of quitting on him a second time, but thrilled at the opportunity to work with kids like me. As I explained to Bill the opportunity, he jumped out of his chair, walked around his desk and gave me a bear hug. "You have to do it!" he exclaimed.

What did I ever do to deserve a friend like this?

I spent a wonderful two years as executive director of a Big Brothers Big Sisters agency. I again mentored an at-risk youth, and I loved it. I learned a tremendous amount about fundraising and the administrative side of running a nonprofit that specialized in relationships and mentoring. From there I returned to the construction equipment world running a sales team for a Caterpillar dealer, then worked for a family-owned general contractor.

I greatly enjoyed being a deacon in a large Presbyterian church. Serving the church family provided a way for me to help others, and I was still searching how to turn my efforts to help veterans. During my third year, the church made me the Chairman of the Board of Deacons. The deacons met once a month on a Tuesday night and the meetings started at 7:00 p.m. and would run very late—sometimes lasting three hours.

The group did a lot of great things and had some big projects, but individuals in the group tended to get long-winded. When they put me in charge the first thing I did was publish an agenda, and I contacted each person on the agenda and warned them that they had exactly the amount of time on the agenda, and if they went over I would cut them off.

One of the pastors had ten minutes at the start of the meeting, and at the ten-minute mark it became apparent he had no intention of stopping. I got out of my chair and stood at his shoulder (he was at a podium at the front of the room). After another minute, I interrupted him and thanked him, but emphasized we were had to keep to the published agenda. He finished and sat down and we moved on. That night the meeting lasted less than an hour. By the time my year as Chair was completed, our meetings lasted less than 45 minutes. One of the deacons joked that his wife was wondering what he'd been doing all those other years, since he used to come home so late.

I believe it is extremely important to honor others' time. Especially if they are a volunteer for an organization, they have full lives, many of them work long hard hours at their jobs, and they have families they want to go home and see. Being in the Army and away from my family for so long gave me a great appreciation for the preciousness of that time and I will guard a volunteer's time carefully anytime I'm put in a position of influence.

By the way, the Board accomplished every bit of the events and service work they had the previous years, and for years after I would get comments from church members about me being "the guy" that changed the way they met.

One of the deacons and I became good friends. Mike Pompeo was a local businessman, graduate of West Point, and had served in the Army. I originally met him when the Wichita Business Journal ran an article about two former officers forming an aerospace company in Wichita. Mike was one partner, and the article mentioned that the other partner, Brian, was an Army Ranger. I sent Brian a note welcoming him to town and inviting him to lunch as there were few other Rangers in the area. He introduced me in turn to Mike, and Mike and I developed a routine of meeting for breakfast a couple times a month.

One morning Mike told me that he was contemplating a run for Congress, since the seat was vacating. I jokingly told him that the fact he wanted it should automatically disqualify him, then volunteered to help in any way. I ended up co-chairing his veteran committee and traveled around the area talking with various groups to gain their support. Mike got elected, and served several terms. In 2017 he became the Director of the Central Intelligence Agency.

During this time came the worst failure of my life. Jana and I got divorced. Jana was amazingly dedicated as an Army wife and a mother, and she was loved by my soldiers. She helped deliver one of my soldier's babies, he was too squeamish to be in the delivery room, and she was a finalist for Spouse of the Year for the 25th Infantry Division. Without her support, I could never have been a Ranger officer. I thank God that we remain friends and that we are both blessed with wonderful second chances.

Through this difficult time, I chose to focus harder on being a father. Someone clued me in to this new social media fad called My Space and Facebook and recommended that I become friends with my daughters on the platforms, so I could keep tabs on what they were up to. I did and when I had it search my email for potential connections, what do you know — a Ranger popped up, and then another! I began developing a network on Rangers on social media.

In 2003 I received a letter from the Army officially placing me on the US Army Retired Reserves. I also received an invitation from Col. Joe Votel to attend his Change of Command at Fort Benning, Georgia. It was the first time I spent any time with the active duty Rangers since my departure ten years earlier. I was unsure of my welcome, still bearing the sting of my departure but with an overwhelming sense of pride and a yearning to return and be a part of that amazing community if only for a brief few days. When I walked up to the parade ground a young Ranger asked my name, then escorted me to the front row where a seat was reserved in the VIP section with politicians and Generals. I was beyond happy. I was so proud to be there and to see Joe relinquish command of this storied unit.

While there I reconnected with old friends still serving, and old friends who had also returned to Benning to rekindle connections. I learned firsthand from them the stress of nearly two years of continuous combat operations was causing this elite fighting force. Combat was tearing them up physically, mentally, emotionally, professionally, and spiritually.

I made new friends and decided to become active in the US Army Ranger Association. Some of my new friends were fabled heroes of World War II. Len Lomell climbed the cliffs at Pointe du Hoc during the Normandy invasion and virtually single-handedly took out a battery of German defensive guns that were preparing to wreak havoc on the invasion force. Tom Herring cracked me up with funny stories about how once he came ashore all he cared about was finding a spot to relieve his bowels, he didn't care that anyone was shooting at him. Sadly, most of these men have since passed away, but the honor and privilege of becoming friends with them will stay with me forever.

I left the reunion determined to find a way to help. These men and their families sacrifice so much yet our nation seems not to notice. They set aside years of their lives to serve while their contemporaries climb the corporate ladder, and rightfully expect to receive recognition and respect when returning to the civilian community, and when they do, they are welcomed home with rejected resumes and a VA that seems not to care whether they live or die. They are hard pressed to figure out how to adapt to a completely foreign environment, one where everyone seems focused on "Me, first."

I volunteered to help the Ranger Association, to revise and update its website, www.ranger.org. I became one of the Regional Directors. USARA is a great organization that has been very supportive of GallantFew's work, especially through the Darby Project, which is GallantFew's program that focuses on soldiers who served in the 75th Ranger Regiment and who have successfully completed the US Army Ranger School. The shared experience of being a Ranger gives a mentor (a Guide) the moral and emotional authority to help and challenge a young Ranger (future Guide) through the transition process.

My connections grew and I created the first Ranger networking group on LinkedIn. Every platform I joined, first I searched to see if one was already in existence. If so, I joined it. If not, I created it. There were no Ranger groups then, so I created US Army Rangers on LinkedIn. To maintain the integrity of the group, I personally verified the Ranger status of each Ranger joining—which was time-consuming but which I greatly enjoyed. If a Ranger was looking for employment, I asked him to send me his resume. By now in my post-military career, I had a wide variety of experience and had run organizations with over sixty employees and managed sales over eight million. I knew what I liked to see on a resume, and felt that could be valuable assistance.

I immediately identified an issue in the resumes that the military's transition program had helped these Rangers produce. Full of military terms and job descriptions, to an untrained reader it made little sense. One simple example is that of a sniper. A veteran with sniper experience might write that he had responsibility for $1.2 million dollars of gear and equipment, operated with little supervision over large areas of ground, decisively defeated the enemy and furthered the objectives of the US government. A human resources person, let's say at a bank, might look at this and think security guard at best or at worst think they didn't need anyone shot today so thanks for your time and good luck.

The reality is that a sniper is an expert at inventory, because in operating in a small team for extended periods of time, knowing how to plan for use, inventory, and resupply is critical to survival. A sniper has advanced mathematics skills. Calculating the flight path of a bullet and figuring in variables such as wind speed, distance, temperature and more, requires a sharp mind, and although snipers use ballistic computers they are required to back up that technology with their brains, in the event the technology fails. A sniper must be cool under pressure, never losing control.

Are inventory management, mathematics, and self-control important to a financial institution?

That sniper may never make it to an interview if he uses only military terms; his resume will hit the discard pile.

The people teaching these soldiers to write resumes have little to no resume experience themselves. They haven't run businesses, don't have human resources certifications. They are for the most part military retirees or spouses. They are not prepared to help a soldier prepare for his or her own transition.

So, I used my experience to help these Rangers translate the mil-speak into terms a civilian employer would not only understand, but would desire. As I did, I would learn more of their life story and began looking for other resources that would help them. I became the hub of a wheel of Rangers.

I also identified a disturbing trend — isolation. When Rangers run into challenges, their true characteristics emerge. They are rugged individuals, driven to personal excellence and used to achieving their objectives. For many of them, that works and they can transition to successful civilian lives. For a significant number, however, the obstacles prove too great — and I can trace most of the issues to a lack of VA care and over-medication, either prescribed drugs, alcohol abuse, or using illicit drugs. No Ranger wants to admit weakness or defeat to anyone, least of all their Ranger comrades, but the longer they try to go it alone the worse it gets.

Chapter 9

Undeserved Second Chances

While at Big Brothers Big Sisters, I had been nominated and accepted for Leadership Wichita, a Wichita Chamber of Commerce program that over eight full-day sessions acquaint community leaders with local government and other resources. After the program, the class selects a trustee to represent the class and oversee the program for the next few years.

I must have been out of the room when they voted, and I was elected. Little did I know, but two years later I would meet Sara, herself a Leadership Wichita selectee. Never married, at the time Sara worked in Human Resources at Cessna Aircraft. We went on our first date in September 2005 and were married a year later.

Without Sara's support, GallantFew would not exist.

As we settled into our new routine, Sara joked that I seem to come home from work and spend all my evening time on my laptop connecting Rangers. "If only you could make that your job…"

As my experience working with transitioning veterans deepened, it became apparent to me that at least in the Ranger community, connecting a Ranger with another

Ranger in the same local area created conditions for success. With few exceptions, a Ranger will go extraordinary lengths to connect with another Ranger.

I first learned this on a business trip to San Diego in the late 90s. Discussion boards were the social media of the day and I posted on a board on www.armyranger.com that I was headed to California, and if any Rangers were in the area I'd love to link up for a burger and a beer.

A 2nd Ranger Battalion veteran immediately responded. Pete Parker was the Ranger who had put the discussion board together and we worked out a plan to link up. I learned later that he had driven over four hours to spend an hour with me. A major reason for this is in the pre-Facebook world it was much more difficult to connect with someone from a very select population such as Army Rangers.

Nicknamed "quiet professionals," most Rangers don't run around boasting about being Rangers. In fact, most legitimate Rangers will deflect the conversation away when asked to talk about their service. Part of the reason is a lack of knowledge because the person asking can't understand what it's like to be a Ranger, unless they are one — and if they were one, they wouldn't be asking.

I've randomly run into more "posers" — men who claim to be Rangers — than I have legitimate Rangers, to the point where when someone tells me they are a Ranger, unless I stood in formation with them, I immediately doubt their word and credibility until they prove their qualifications. Usually this is a simple process of asking a few questions about the Regiment and their chain of command. I'm not sharing that here because I don't educate posers.

The first one I met was in my first job selling construction equipment. Seeking to create some connection with an individual who was infamous among salesmen for being a blowhard, I took him to lunch. When I asked about his past, he volunteered he was in the Army. When I asked what he did in the Army, he told me he was a Ranger.

I reached into my pocket and placed my Ranger coin on the table between us. His face turned bright red and he didn't pick it up. He stammered that he actually was a medic who took supplies forward to Rangers in Vietnam, but I knew that was a lie, too.

I had a poser pretended to be my friend, Ken Stauss, who had been killed in the crash at the Great Salt Lake. He represented himself to me by describing every bit of Ken's service as his own. He had no idea I knew Ken so well, and when I called him on it he confessed that he had served under Ken in another, non-Ranger unit and respected him so much that he took his story as his own. I was furious and it's a good thing we were on the phone and not face-to-face.

One more poser story — I had a man represent himself as a retired officer, working with the Cleveland Clinic to furnish reconstructive surgery to severely maimed soldiers, and he had discovered my connection with Mike Schlitz (more on this later). He got Mike and I on a conference call, and talked about how he wanted us to help him pitch his program to the Clinic. While on the phone, he bragged about his awards for valor (Distinguished Service Cross, second only to the Medal of Honor) and others. He dropped names of officers I knew, stating that they wouldn't admit it due to the classified operations they'd participated in together. He also claimed to have been part of Special Forces Operational Detachment—Delta, known as Delta Force. No member of that unit that I know would ever admit to someone they didn't know and trust that they served in that unit, so my alarm bells were ringing. Oh, and by the way, his records were "classified."

He sent me an email with a link to his LinkedIn profile, and I saw that his email address included "mg" before his name in his yahoo email address. I asked if he was a retired Major General, and he responded that yes, he was.

I called him out. No two-star general I know would *ever* put their military rank in a yahoo email address. Sure-fire giveaway. No personnel records are classified and no awards are classified (although the circumstances might be). If anyone ever claims as such, you just met a poser.

Once I enjoyed exposing these posers to social media, then we had a misfire. During a check of Ranger School records, we got a false negative—"not a Ranger." It was a mistake. One of our admins removed the Ranger from a social media group and it created a firestorm, embarrassing the Ranger and embarrassing the admin (also a Ranger). I've also worked with soldiers with traumatic brain injuries who have difficulty knowing fact from fiction.

I decided to be known as someone who helps, not someone who hammers. I refer real posers to law enforcement if they are seeking to profit from their lies, and I tell them personally to clean up their act or I won't help them—but I don't want a soldier to not approach me for fear of being publicly embarrassed for exaggerating war stories.

Now—where was I? See, posers can be a real distraction from the real work.

As I continued to connect Rangers and help them as they transitioned between jobs, processed combat experiences, or battled substance abuse, I saw an announcement that the National Institute of Mental Health wanted proposals for unique, innovative ideas for treating PTS. I put together a paper proposing a "Ranger Buddy" program, where Rangers would be connected with a local Ranger buddy to help them through their experiences. I reasoned that if it worked for Rangers as a test population, the Army could readily expand it to the rest of the force, ala Abrams' Charter.

A few months later I received a "thanks but no thanks" letter.

Now I had started to put parameters around this program. I refined the paper and sent it to several Rangers I knew and highly respected, one of which was newly retired Command Sgt. Maj. Mike Hall. Gen. McChrystal referred to Mike in his book as the finest soldier he ever knew, and Mike and I had served together in the 1st Ranger Battalion, recently reconnecting on Facebook of all things. Mike responded right away that he thought the idea had merit and he'd like to be involved. Then a month later he sent me a message that McChrystal had requested he come out of retirement and serve as his senior enlisted advisor in

Afghanistan, so he'd have to postpone his involvement. A year later he retired again, and has been part of the GallantFew Board of Directors ever since.

At Ranger Rendezvous in 2009 I approached several leaders in the existing Ranger associations about my idea of creating a process to connect Rangers leaving active duty, with Rangers established in their civilian careers. There were several reasons they gave me for not being interested — this was not in their mission, they don't have the resources to take it on, they are more interested in the reunion aspect than taking on a service agenda. I continued developing the plan.

In October, 2009, auto executives from the largest domestic manufacturers flew their corporate aircraft to DC to testify to Congress how badly their business was doing, and were excoriated in the press for the excess demonstrated in flying private jets. Overnight, private aviation became persona non-grata and customers cancelled orders. This had immediate and devastating impact on the Wichita economy. The original home of Cessna, Beechcraft, and Learjet, since World War II the aircraft industry had played a major role in Wichita. Boeing Aircraft had a large facility, and the President of the United States' own aircraft, Air Force One, was outfitted with sensitive communications electronics there.

Orders for aircraft dried up. The family-owned business I worked for depended greatly on the aircraft industry for work and found itself suddenly with more people than work. After a Veterans Day weekend where Sara and I went to an event in Arkansas, representing the US Army Ranger

Association, I returned to work to discover I'd been laid off. Two weeks earlier my boss had told me I'd receive a bonus; that's how quickly the economy there turned.

I swore I'd never work for anyone again. I'd figure out a way to make my own way.

I sat at dinner with Bill, my best friend and mentor and whom I'd quit on twice before. I told him of my experiences connecting Rangers and how I believed the government transition system was inadequate and that I felt I had something that could make a difference.

He smiled, pulled out his wallet and removed a hundred dollar bill he kept hidden behind his driver's license. He slid it across to table to me and said simply, "Get started."

I met monthly for breakfast with a high school buddy who was now an attorney, and I found myself telling him about the issues I was observing with veteran transition and my concept of one-on-one mentoring. He suggested that I form a 501 (c) 3 nonprofit and offered his law firm's support to make this happen. We submitted the required documentation in January, 2010 and that summer received the approval from the IRS. GallantFew, Inc. was in business.

I've been asked how the name GallantFew came about. As I researched options after being laid off from my job, I met an Army veteran and local businessman named Duke Naipohn. Duke was a former airborne trooper and after the military had pursued education in respiratory therapy. While working in a hospital, he cared for an elderly man who was dying of respiratory disease. Somehow Duke found out that this man was a World War II paratrooper, and, you know it, they became best friends for the few remaining days of his life. Following his death, the daughter of the man gave Duke a bracelet that bore an inscription on the inside "from a grateful nation to a gallant few." Duke looked at me and said simply, "There's the name."

Michael Schlitz

Around the same time, I got a message on social media from an active duty Sgt. 1st Class named Michael Schlitz. He wanted to join one of the Ranger groups I had established and I shot a note asking for his qualifications. He answered back that he had served several years as a Ranger Instructor but that he never got to serve in the Ranger Regiment. He had been accepted to start the review and evaluation process for potential assignment when he was catastrophically injured in an Improvised Explosive Devise (IED) attack in Iraq.

Composed of two 155mm artillery shells and a propane tank, it tore through his soft-skinned vehicle like a hot knife through butter. The blast threw Mike, now on fire, clear of the vehicle, also burning. He landed on the ground, got up not realizing he was on fire, and ran towards his vehicle and the three soldiers trapped inside. Two had been killed instantly but Mike could hear the screams of the third, then the flames overpowered him and he fell face first to the ground, burning alive. His soldiers put into place their well-rehearsed drills, sprayed him with a fire extinguisher and within ten minutes he was on a medevac bird headed to the hospital.

Mike was undergoing treatment for his burns at the Army's medical center in San Antonio, Texas. Now why do they send burn patients to the hottest city in the USA? Seems like Michigan would be a better place – but that might be too much common sense. I've traveled extensively with Mike and heat affects him badly, he has no sweat glands left and his skin is completely covered with scars. Mike's injuries were so severe that both of his hands had to be amputated mid-forearm. He lost his lips, nose, ears, and eyelids to the flame.

His mother, Robbi, is an absolute angel. She gave up her life, her career, income, even health insurance to become Mike's full-time caregiver—and Mike required full-time care. Mike couldn't feed himself, couldn't use the bathroom without major assistance, and he suffered from such depression that he considered suicide. He tells people now the reason he didn't take his own life is he was never alone, and he didn't have any hands.

I've learned more about a human state of grace from Mike Schlitz than from all the other humans I've ever met.

That summer, as Central Region Director for the US Army Ranger Association, I was responsible for planning and conducting the Annual Ranger Muster (ARM) which was to be held in San Antonio. I quickly enlisted Mike's help since he was local to the area, and he was eager to have a mission. I traveled to San Antonio prior to the ARM, and I dropped Mike a note that I'd be there, there wouldn't be much time to meet, but perhaps we could grab lunch before I had to make my flight. He told me he would pick us up (I was there with one of the first GallantFew board members, Alan McNiel) and we'd eat on the way to the airport.

When he said pick us up, he meant his mom would be driving. Years later I got to ride in a Polaris with Mike driving. It's an experience to ride with a guy who can't see, who doesn't have hands, and who likes to go fast—and it was at night.

Mike and his mom pulled up to the curb and Mike got out of the passenger door and I was stunned. Mike had yet to go through about half of the 80 plus surgeries, mostly through Operation Mend at UCLA that does amazing work rebuilding these physically wrecked soldiers, but it wasn't the fact that before me was a man who was one massive scar tissue, with a hole where his nose should be, no ears, and hooks in the place of his hands, that stunned me. What was stunning was how he smiled, and extended his hook for me to shake. I pushed his hook aside and gave him a hug.

The only restaurant between where he picked us up and the airport was Hooters. You want attention from the waitresses at Hooters? Go there with Mike Schlitz.

Thinking I would amuse Mike, I told him a joke I'd heard once, about a pirate that had a hook on one hand only one eye. 'The joke went something like this: "Hey pirate, how'd you get your hook?" "Well, laddie, a shark got me." "Well, how'd you lose your eye?" "Well, laddie, a bird pooped in it." "A bird?" "Yes, laddie, it was the first day with me hook!" Mike laughed like it was the funniest thing he'd ever heard. Six months later we were at a veteran event, and someone came up to Mike. "Hey, Mike, have you heard the story about the pirate?" Mike listened intently and laughed like he'd never heard the joke before. I've seen this happen at least a half dozen times. Mike is a special kind of human. He doesn't cut people off by saying, "Yup, sure have..." — he makes them feel special.

I've stood with him at events and he is a celebrity magnet. They are drawn to him and they ooh and ah over his injuries and they ignore whoever might be standing nearby. I've seen Mike interrupt them and introduce the veteran next to him, stating something like, "You might not see it, but this veteran broke his back and can never parachute again." He deflects attention away from himself and onto others.

I was with Mike when for the third time the VA asked for more evidence of his injuries to justify his 100% disability rating. If a visibly, horrifically wounded soldier like Mike has difficulty proving his injuries to the VA, what chance do the rest of us have? Thank God for organizations like the Gary Sinise Foundation and the Carrington Charitable Foundation. They and others like them build adaptive homes for severely wounded soldiers because our inept government can't or won't.

Mike is a rock star, and he has been one of my best friends and advisors ever since we met.

Karl Monger, Bill Cooper, and Mike Schlitz, 2012

The Big Three

From the beginning, GallantFew set out to create grass-roots, effective ways to solve the "Big Three."

There are three major issues that define the veteran "problem." They are unemployment, homelessness, and suicide. It seems simple to say that first you lose your job, then you lose your home, then you lose hope but it's much more nuanced and complicated, and each of the five points of the GallantFew STAR address these areas.

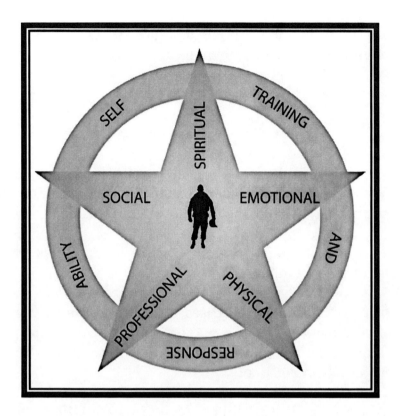

First let's look at the facts.

Suicide

The Veterans Administration published a study in July, 2016 of veteran suicides over a period of time from 1979 to 2014.

(https://www.va.gov/opa/pressrel/pressrelease.cfm?id=2801)

It concluded that on average, twenty veterans a day committed suicide. That's 7,300 a year; in fifteen years since 9/11 that exceeds 109,500 souls.

Let that sink in a moment. Since 9/11/2001, 109,500 veterans have taken their own lives.

109,500.

When is the last time you saw that on the news? I'll bet never.

What priority does our government, does our country, place on veteran mental health and veteran suicide?

Let's put it in perspective with other issues most Americans find familiar:

The CDC (Center for Disease Control) says nearly 28 million Americans suffer from heart disease, and the government budget allocates $818 billion to combat it. That's about $30,000 per patient.

Approximately 1.2 million Americans are HIV positive. The government allocates $27 million to treat patients and combat the disease; that's over $22,000 per patient. 6,700 HIV positive individuals die from the disease every year.

There are 21 million veterans in this country, and the government allocates $7 million to veteran mental health. That works out to $333 per veteran. 7,300 veterans die by their own hand every year.

The dollars speak for themselves. Veteran mental health is not a priority in our country. I believe a key reason for this is that the veteran community does not forcefully advocate for itself, and before social media made it possible to share experiences far and wide, the individual veteran issues were exactly that—individual. Now, however, we know the universality of these issues, and it's time we made our voices heard. More on that later in Actions Steps, at the end of the book.

The VA also says in that study that in 2014 over sixty percent of the veterans who took their own lives were over the age of 50. Since 2001, the risk of suicide for a veteran is 21% greater than that of a non-veteran. Why are veterans killing themselves at rates so much greater? Why are more than half over fifty?

The number one reason for any suicide, particularly a veteran's, is loss of hope and purpose. It takes years to get to the point of desperation and hopelessness, thus the older statistic.

Unemployment

Unemployment remains an issue that also receives rare if any news attention, and it seems the government does its level best to hide the true issue. On February 3, 2017, the Department of Labor trumpeted through their e-newsletter that veteran unemployment rate at "4.5% remains below the national average." I've been tracking the actual numbers since January, 2011. I keep them in a spreadsheet so I can run monthly numbers and annual averages. The truth paints a different story. It is true that overall veteran unemployment for January 2017 was 4.5%, and the national average for nonveterans was 5%. In fact, over the last twelve months, a veteran was nearly 10% less likely to experience unemployment than a nonveteran.

However, a Desert Storm veteran (the government calls that conflict Gulf War I) has an extremely low unemployment rate—in January, 2017, it was 3.4%. That's 32% less than a nonveteran! But when you look at the post 9/11 veteran (the government calls that Gulf War II), the rate is 6.3%, 26% worse than a nonveteran and nearly double the older veteran population. Some will say, "But this is because these are young veterans, and they are in school." If they are in school, they aren't drawing unemployment so they won't figure into the calculation. Over the last twelve months, a post-9/11 veteran is 21% more likely to be unemployed than a nonveteran, while a Desert Storm veteran is 30% less likely to be unemployed than a nonveteran.

The government hides the problem by using the great employment record of the Desert Stormies to mask the problem with the veterans that have been volunteering to fight while we are a nation at war. The first step to solving a problem is correctly identifying the problem, and Houston, we have a problem. Veterans under the age of 35 are doing significantly worse than anyone else.

This also tells me that the older veterans have great jobs and careers; that they have figured it out and stabilized. Further, the US Census tells us that 6% of the US population is made up of veterans, and that 0.5% are veterans under the age of 35. In 2016 we surpassed 15 years at war, 35-15 = 20, so these are all post-9/11 veterans. One-half of a percent of the population are post-9/11 veterans under the age of 35. Given that these numbers average out over most major population areas, a community of 100,000 people will have 6,000 veterans, 300 of whom are post-9/11, under the age of 35. That means that 5,500 veterans have great jobs and are established in their communities, and are in ideal position to mentor, guide, network, and connect with 300 veterans who might be struggling. That is a greater than 10 to 1 ratio, and that's a lot of numbers for a Ranger to process.

But here's the key takeaway: In every community across the country, *there are potentially ten veterans in position to help one.*

Okay, so not all the older veterans are able to help, some are well past retirement, disabled, perhaps unemployed. Not all the young veterans need help. It's still ten to one. Those are great odds; we just need to leverage them to benefit our communities.

Homelessness

The third of the big three: homelessness. The VA has put a bullseye on the homeless issue, and I believe it's because it's the most visible. A disheveled veteran at a street corner begging for money is an embarrassment to our nation, our communities, and especially to the VA.

But throwing massive dollars at homelessness is addressing the symptom, not the cause. If a veteran has a job, if he or she has great mental health care, if the veteran is connected in the community, they are not going to be homeless. Responding to a homeless veteran is a reactive move, but necessary, and the VA claims to be doing well. I suspect local community efforts to resolve the issue is making a greater impact than the federal response—but we still must get to the root cause, and put programs in place to *prevent* that veteran from ever getting to the point of homelessness.

I recently helped a veteran on the verge of becoming homeless. He is an older veteran, having served decades earlier. Many organizations established to help veterans won't consider his needs because his service predates 9/11. To me, he volunteered, he offered his life for our freedom and now that he's in a pinch he deserves some help. He would soon be homeless because he'd gone through a medical issue that prevented him from working. As he burned through his savings he moved from house to long-term rental, eventually moving in with a relative in a tiny apartment. The relative was moving out of the area soon, and with no income and no other place to stay, this veteran and his wife were planning on sleeping in their car. They had already given their precious dogs to foster families in preparation.

I sent him to the local VA and the state veteran commission. They referred him to the US Department of Housing and Urban Development. HUD told him that they couldn't help him, because he wasn't yet homeless. Once he moved into his car, then he could apply — and they told him to expect a processing and wait time of about two weeks after applying.

This is solving veteran homelessness?

I learned about him because a local resident had heard about GallantFew and the monthly veteran breakfast. I had a conversation that I know was difficult for the veteran as we covered in detail why and how he was in this situation. We agreed he was responsible for his success and together we put together a plan. Ultimately the veteran didn't have to live in his car, and maybe that's our government's plan. Make it too difficult to get the help you need and you'll figure it out on your own.

Without the visibility of the local informal veteran group and the referral what would have been the outcome? I don't believe this veteran was suicidal but it gives credence to the VA's own study, that older veterans are more likely to end their lives than younger ones. They simply exhaust all their options until they have nothing left.

This Is Revolutionary

GallantFew's tagline is "Revolutionary Veteran Support Network." What makes it revolutionary is our preventive, proactive approach: we seek to provide every person leaving active duty with a Guide who has a transitional AAR – gold nuggets of information that can change the course of a life and the moral and emotional authority earned through shared service that gives the credibility which provides the foundation for the relationship. If we can intercept them, keep them on azimuth by preventing isolation, and by sharing the lessons learned – our transition AAR – then I believe the problems of suicide, unemployment, and homelessness will not be manifested.

Chapter 10

Staying on Azimuth

Transition is like night land navigation, old-school.

For those of you who haven't had the fun, here's how it works: lie on the ground and cover yourself with a poncho or other cloth that blocks light and plot two points on the map: where you are now, and where you desire to go. Since it's going to be dark, you won't be able to use a far-off reference point, like a noticeable tree or hilltop as an easy reference. Instead, you are going to have to rely on your ability to walk in a straight line while counting your paces to determine how far you've traveled.

Using a protractor, draw a line between the two points you plotted on the map. Ensuring you figure in the declination factor (degree of difference between grid north and magnetic north), determine the azimuth from your present location to your desired location.

Measure the distance you are to travel, and remember it.

Open your lensatic compass, orient it to north and dial in the azimuth you figured earlier. If the glow-in-the-dark lines or dots aren't glowing now, get a different compass or use a light to charge them up (and hope the charge sticks). Turn the compass until the north arrow glowing line lines up with the glowing azimuth line. As long as the north arrow stays aligned and you walk in the direction the compass is pointing, you will stay on azimuth.

Know your pace count. I know from experience that at a normal walk, 120 paces (or 60 with my left foot), equals 100 meters. Now is a good time to ensure you also have your pace count cord, a cord containing sliding beads. Every 100 meters you will slide a bead down so you don't lose count.

Start walking.

If it's truly dark, and if you are walking on uneven ground or better yet — through a forest or swamp — you will soon find that it's hard to keep walking in a straight line for very far. You'll encounter a stump, or a water-filled ditch, or walk into spider-web, or face a large tree that you simply can't walk straight through. Now turn 90 degrees to the right or left, walk the number of steps required to clear the obstacle, pick up your original pace count and azimuth and walk straight again. Notice you have moved your azimuth to a parallel azimuth the number of steps you took to clear the obstacle. Once it's safe to return to the original azimuth, face the opposite direction you did earlier and walk the exact same number of steps back. Theoretically this action places you back on the exact azimuth with which you started.

I learned to alternate the direction I took around obstacles. If I went right the first time, I went left the next. This evened out any tendencies I might have to step unevenly or to miscount.

If you are precise about this, when you reach the pace count that corresponds with the distance you desire to travel, you should be right where you planned to be.

What happens, though, if you go around a tree twice or three times to the right instead of alternating? If a spider web freaks you out so much that you do the herky-jerky, trying to clear the spider web and any associated critters from your clothing. When you regain your composure, you have no idea how far off your azimuth you went. What happens is, you continue your azimuth as best you can, and when your pace count tells you you're near, you stop and look for your spot.

If you've only gone one or two football fields in distance, odds are good you'll quickly find your objective, because a couple of degrees at that distance isn't significant. If, however, you walk ten miles, now that's a different story. You'll end up so far away from your original objective that you may never find it.

You're probably asking yourself at this point, "Great lesson in land navigation but how does that relate to transition?" Well, I believe there is a direct correlation.

When you go through ACAP, or TAP, or whatever your branch of the military calls transition training, all that information becomes your azimuth. That is what dictates the path you are going to attempt to follow as you transition.

Odds are you are not going to remain in the same community where you were last stationed. You are going to move to a location for a job opportunity, for education, perhaps you are going home, or moving to a place you've always dreamed of living. Even if you go home, you are now years removed from the community. Your friends are either working hourly jobs or if they went to college, might just now be graduating and looking for work themselves. In any event, the newness of the community is the equivalent of nighttime. You have no idea what you are getting into until you step into it.

Spider webs, big trees, ditches—these are the small obstacles you will encounter. You don't get a job right away, or you have difficulty meeting new friends, money is tight, school is frustrating, the VA is unresponsive, you experience survivor's guilt, you may drink as a form of self-medication—there are a million obstacles you may encounter. Each of these obstacles has the potential to take you a degree or two off your transition azimuth.

How then do you get back and stay on azimuth? What if you had a map that had marked on it the exact location of snakes, big spiders, dangerous holes, as well as great resupply points and rest stops? Well, that map exists in every community, and should be readily available to every transitioning veteran—and that map is another veteran, one who is a "transitional generation ahead." In other words, he or she left the military a few years ago, completed their schooling, got a job, became established in the community, has a network of friends and business associates, belongs to a Rotary Club or a Chamber of Commerce or a faith-based group (or multiples of these). This person has walked the transition azimuth, strayed from it, bounced back, maybe figured out a new azimuth, and in all likelihood, would be more than happy to share how they did it with you!

GallantFew calls this veteran a "Guide." This Guide has a transition After Action Review (AAR), and you need the lessons already learned. GallantFew calls the transitioning veteran a "Future Guide," because once you've accumulated your own transition AAR, we want you to share it with someone following in your footsteps.

Imagine if there was an effort in every community in the country to identify veterans willing to be Guides to Future Guides. Imagine if every veteran returning to or moving to a community was greeted by a veteran just like him or her: the same branch of the service, the same military skill-set, perhaps the same deployments or even the same or similar injuries. I believe this network of veterans would grow to become a solid, important part of every community. Instead of veterans fighting unemployment, self-medicating, struggling to deal with experiences and frustrated with the VA, they would be networked, in supportive friendships that challenge, motivate and inspire.

How many families might stay together, how many new businesses would start, how many suicides would be averted?

In Chapter One, A Tale of Two Rangers, I talked about Steven Barber and his new State Farm Insurance Agency. Steven directly attributes his ability to take a risk on the career of his dreams because of the simple fact that he had an instant network, whereas before he might have taken the first job offered because he had a family to support, and no military retirement to provide a cushion.

A few years ago, two Ranger veterans decided on a whim to see how far they could travel from Denver, Colorado with a hundred bucks and what they could fit in their backpacks. They published a documentary of the journey, called *Nomadic Veterans* and it is available through Amazon and Netflix. Early in their journey, they tried to do it alone and ended up sleeping in a field, hunkering down outside a storage building in a tornado, and didn't get very far. As soon as they tapped into their veteran network, suddenly they were getting rides; people bought them meals, even plane tickets overseas. Their journey was amazing and their biggest lesson? That doing it alone got them nowhere but as soon as they turned to their network, things began to happen.

As of September 2016, our country had been at war for over fifteen long years. I work with veterans who have been off azimuth, trying to find their way alone for years and years. They have burned through relationships, money, they are isolated, frustrated, they drink hard, and they don't have much hope that their situation will ever change. They have no sense of purpose and believe the best part of their life is now behind them. They are so far off azimuth that it becomes a significant emotional event for everyone involved to establish a new azimuth to get back on track. Sometimes it involves swallowing their pride and admitting they need professional substance abuse help, or post-traumatic stress support. The resources exist to help them start from right now to change their lives for the better.

You know by now that I'm a former Army officer, infantryman, paratrooper, and Ranger. When I have a conversation with an infantryman, paratrooper, or Ranger, I can challenge them to overcome their obstacles because that soldier and I have been through the same or very similar experiences. We trained at the same places; we probably know many of the same people. I have the emotional or moral authority to challenge them. It's less so when I work with a Marine veteran. The Marine can rightly say, "You haven't been through what I've been through, so how do you know?" That's where a Marine veteran, transitionally a generation ahead, can get eyeball to eyeball and encourage, support, and challenge the Marine to start making changes.

More on this in Chapter 11, The Guide and the AAR.

Chapter 11

The Guide and the AAR

"We few, we happy few, we band of brothers…"
~ Shakespeare

"They need a little bit of light, to be told it will be OK. I did it, you can do it."
~ Anonymous Ranger veteran

"I wish this was available when I got out."
~ Every veteran I talk with

"Let my hindsight be your foresight."
~ Unknown

"The connection is the cure."
~ Josh Collins

I have done very dangerous things in the military. I have placed my life and the lives of my men in danger. I could do these things, and do them with the confidence that we would emerge unscathed, because I had learned from generations of soldiers who had gone before me. Their lessons gained through blood, sweat, tears, and death were passed down to me so I wouldn't have to learn them in the same painful ways.

Nothing is done in the military without an After Action Review (AAR). Training exercises, combat operations are all picked apart in an effort to capture the things done right (so they can be shared and repeated) and the things that didn't work (so they can be shared and avoided). The Army even has a command at Fort Leavenworth called The Center for Army Lessons Learned — yet the concept of an AAR relating to transition is nowhere to be found in any of the transition programs offered. Every person leaving active duty gets to learn the same lessons without the benefit of the experiences of thousands of other veterans just like them.

One of the worst situations to be caught in is an ambush. The enemy has carefully selected a kill zone, ensured his deadliest weapons are situated to provide devastating fire throughout the kill zone, and waited patiently for the unsuspecting prey to walk into certain death.

To freeze or to try and take cover in a kill zone is certain death. The only chance for survival is to charge headlong, recklessly straight into the oncoming fire. It's not a good chance, but it's a chance.

How do you suppose we know this is a chance?

It's because someone did it, survived, and shared the story so others could benefit from the knowledge. Now every soldier learns "react to ambush" as a basic battle drill, and units constantly rehearse their response.

Another dangerous activity soldiers do is jump out of aircraft in flight. Before every jump, US Army regulations require all the soldiers jumping to participate in Sustained

Airborne Training, or SAT for short. During SAT, the jumpmasters (soldiers who have advanced paratrooper training and are charged with ensuring that the jumpers safely exit the aircraft and are prepared for any contingency that might arise) march the jumpers to an open field and run them through contingency drills. Such contingencies include the possibility of landing in a tree, in water, on powerlines, having a canopy that fails to deploy, or even landing on another jumper's parachute.

The jumpers stand in extended formation on an open field and the jumpmasters rehearse the jump, going through all the jump commands and the jumpers mimic the required actions, all of which were drilled into the soldier during Basic Airborne Training at Fort Benning.

"Jumpers hit it!" All jumpers hop a few inches in the air and lock their bodies tight, feet and knees together, arms locked in front, down to the sides at elbows bent, forearms pointing straight forward. To the observer, it looks like everyone is hunching over, holding a small precious box. Everyone chants in unison, "One thousand, two thousand, three thousand, four thousand." On the count of four, one should feel the opening shock. Sometimes the jumpmaster yells, "Five thousand, six thousand!" and the jumpers respond by pretending to deploy reserves. Rangers jump at 800 feet above ground level, and usually at night. It takes about eight seconds for 200 pounds to fall 800 feet, so there is zero room for error.

On the command, "Check canopy," jumpers stand erect and extend their arms up, grasping the risers to the invisible parachute, to ensure there is no malfunction, and the jumpmasters call out potential malfunctions.

"Trees!" Jumpers fold their arms over their faces, peering beneath their armpits to prevent tree branches from poking eyes.

"Power line!" Jumpers pretend rocking the parachute risers back and forth, to increase the possibility of slipping through the power lines rather than getting snagged and possibly electrocuted.

"Landing on another jumper's canopy!" A hundred jumpers jog in place, feeling slightly ridiculous pretending to run off another jumper's chute. The purpose of these drills is to cement into the mind the actions you must take to increase your odds of survival should the worst occur, and these drills work.

In 1991 the Ranger Regiment conducted a change of command. Occurring roughly every two years, the Regiment change of command caps a week of "Ranger-ific" events and competitions. Called Ranger Rendezvous, Ranger veterans from around the world return to Fort Benning to socialize, tell stories, and reconnect with the experience of being an active duty Ranger. All the Regiment's subordinate units (that are not deployed) conduct a rare daylight parachute assault. Timed to the exact second, aircraft from the east and west coasts converge over Fryar Drop Zone at Fort Benning. Few units in the United States military are capable of conducting such an intricate and massive combat insertion, and the Regiment uses the change of command as a means of ensuring this proficiency stays at a high level. Mass-tac is short for mass tactical — "mass" meaning a lot of jumpers in the air, and "tactical" meaning they bring all their combat gear with them. The Regiment had demonstrated the lethal effectiveness of this proficiency two years earlier when it had jumped into history during Operation Just Cause in Panama, conducting simultaneous parachute assaults onto Torrijos/Tocumen International Airport and Rio Hato Airfield. The Rangers captured over a thousand enemy prisoners of war and nearly 20,000 weapons.

Following the Regimental mass-tac, Rangers assemble and spend the week competing in athletic and shooting competitions culminating in transfer of the Regiment's colors from the outgoing to the incoming commander.

As I exited a C141 jet over Fryar, everything seemed normal. On a mass-tac, jumpers use a non-steerable parachute to minimize the risk of mid-air collisions. Jumpers are connected to a steel cable inside the aircraft by a static line. On the command "Go!" the line of jumpers (called a stick) walks rapidly towards the doors at the rear of the plane, attempting to keep about a one-second interval behind the jumper ahead. The jumpmasters (two — one at each door) control the flow so that ideally jumpers exit alternatingly, first the right, then the left door at half second intervals. Depending on number of jumpers and length of the drop zone, the jumpers might surge towards the door, as no one wants to miss the drop zone and end up hanging from a tree.

Exiting a jet is different than exiting a propeller driven plane. One must grab the doors of a prop plane and jump up and out to clear the side of the plane. On a jet, however, one just steps out and, if done correctly, the slip-stream creates a gentle slide as the static line unfurls from the parachute worn on jumper's back. Still connected to the plane, the jumper is horizontal to the ground until the opening shock of the canopy, when the force swings the jumper to the other horizontal, then back until the jumper is suspended under canopy, descending to the ground. I have many memorable pictures in my mind of the belly of the plane, suspended in time and space, floating away from me as my chute deploys.

On this day, however, before I could enjoy the view, I landed flat on my back on a parachute.

When I was at the 1st Ranger Battalion, there was a plaque inside the front door of the battalion headquarters. It honored and memorialized Rangers killed in training. With one exception, all the fatalities to this date were listed in pairs—a dozen Rangers killed when their parachutes became entangled. Military parachutes are designed to support the load of one paratrooper with equipment. When two jumpers become entangled, one of the parachutes will collapse leaving the other overloaded and both parachutists will plummet to certain injury and possible death.

As I sat up on this big green nylon canopy, I noticed that there was a pool of green parachute cord (suspension lines connecting my risers to the parachute) lying on my lap. It looked like someone had dumped a five-gallon bucket of green spaghetti noodles on me. Then I saw that my parachute was sliding off the side of the canopy on which I sat, much like a silk sheet would slowly slip off of a bed. Reaching with my left hand, I grabbed as many of the suspension lines as I could and flung them to my left, then grabbed the rest with my right hand and flung them to the right, and rose to my feet. The sensation was like being in a kid's moon walk, the air below creating enough pressure on the canopy that I could stand. Without thinking, I ran to the spot where my canopy had now disappeared over the side. I jumped, passed my parachute in the air, it caught air and re-inflated, and I pulled hard on a riser, spilling air from the canopy so I would move away from the other jumper but increasing my already too-fast rate of descent.

Looking over my shoulder, I saw the other jumper, but all I remember are eyes so wide open they looked like comic bulging eyes. Looking towards the ground, I saw a pile of sand, and I pulled hard on the riser, willing myself into the sand. I hit so hard it knocked the air out of me—but thankfully it was sand. Had I hit on harder ground I know I would have broken something. The adrenaline hit me and I rolled over, pushing myself up as I gagged and wretched. A one-star general I didn't recognize, who must have jumped on the pass before me, jogged by with his gear, headed for the assembly area. He looked at me and said something like "Hooah! Good training!" and kept on going.

"Thanks for the help, sir—I'm okay."

I survived that near-catastrophe because it had been drilled into my head through dozens of SAT sessions. I had no time to think about what to do; I barely had time to react as it was. Once the jumper below me touched the ground, the force holding the air pressure in the parachute would have been gone. The canopy would have instantly collapsed, and I would have fallen more than thirty feet and perhaps landing atop the other jumper.

The point of this story is that when you are in the military, you have access to a near-limitless knowledge base of lessons learned. Every operation I participated in, concluded with an After Action Review, or AAR. The AAR collected at a minimum three positives and three things to improve for each critical area of the event.

When I was on active duty, the Army used eight battlefield operating systems to provide a framework for evaluating operations. They were:

- Intelligence
- Maneuver
- Fire Support
- Mobility/Countermobility/Survivability
- Air Defense
- Combat Service Support
- Command and Control

Everything one does in the military — training, combat operations — is pulled apart in an AAR using the operating systems framework to guide the review.

The AAR is central to the learning process in the military and aims at identifying not only things needing improvement, but also things done well. Feedback is the bedrock of military training, and is the greatest gap in support for a transitioning veteran.

Where are the transition after action reviews? Has a veteran ever been asked to document his or her transition lessons learned?

Until now there has been no "Transition AAR." GallantFew seeks to create this in two ways. First, by collecting and sharing our collective transition experiences. Second, by the simple relationship of a local previously transitioned veteran mentoring (we call it being a Guide) a veteran undergoing transition.

It is key that it is local. The transitional AAR of an Army veteran in Dallas might not be helpful to a Marine Corps veteran in Los Angeles—but the AAR of that successfully transitioned Marine Corps veteran can be priceless (and life-changing) to a Marine just returning to LA.

I believe there are five transition operating systems. They are all areas of "functional fitness." Every veteran who has transitioned has lessons that have been learned in each of these functional areas—both things that worked and things to avoid. I'll explore each of them in a subsequent chapter.

I use the term "fitness" instead of "wellness" because wellness can carry connotations of illness and is often tied to mental illness. If you're being trained or coached on wellness, it's because there's a deficiency. High achieving soldiers don't like to think they are deficient, but they are all driven to improve their levels of fitness.

The areas are:

- Functional Physical Fitness
- Functional Social Fitness
- Functional Professional Fitness
- Functional Spiritual Fitness
- Functional Emotional Fitness

By keeping these functional fitness areas in mind, we can both capture lessons learned and intentionally plan how to improve or increase our level of fitness in each functional area.

A veteran who has previously successfully transitioned is a veteran who is a "transitional generation ahead." This veteran made it through school, landed a job, dealt with the VA, worked through relationships, made connections, joined organizations such as Rotary or Chambers of Commerce, has golfing buddies, perhaps a church fellowship group—and many times when you ask this veteran how it was done, the veteran will point to a local veteran they were fortunate enough to bump into. This person became a trusted Guide, but it was by accident. How many veterans don't get this lucky, and struggle—living lives that are a shadow of what they could or should have been, just like Mike who I told you about in Chapter One. I could have been a Mike, but for my buddy, Bill.

Perhaps the greatest lesson is this: The sense of loss, reduced self-confidence, struggling to find a sense of purpose, and the tendency to self-medicate with alcohol are nearly universally experienced by transitioning veterans. I've had combat hardened special operations soldiers choke back tears with the simple knowledge that what they were experiencing was normal to soldiers like them, and that knowledge helped them better to accept it and caused them to become eager to learn ways to improve their fitness in that area. Previously they had thought something was wrong with them and it made their transition struggle more acute.

Our nation's veteran care system tells severely disabled soldiers they are broken, and pays them handsomely to go away. The fact that the VA requires nothing in return for a 100% disability rating destroys the very sense of self-esteem so important to emotional fitness and health.

A Guide arms a soldier with knowledge, and with that knowledge comes confidence that empowers the veteran to know the enemy so the enemy can be defeated. It's easy to begin to compromise — "I don't need to get up and do PT today, I'm tired, or it's cold outside." A compromise becomes a reason, then an excuse, then thoughts turn negative as this question arises, "Why can't I get up and exercise?" (More on this in Emotional Fitness, Chapter 13) Then habits change from what kept one mission ready to one that softens the midsection and the mind and holds one captive to a lower level of performance. Simply knowing that can happen, may arm a veteran with the motivation to avoid the insidious process.

Just as in surviving an ambush, or in landing on another jumper's canopy, there are many known situations that a veteran will most likely encounter, but the Veterans Administration doesn't deal with them until they occur. These situations may be geographically specific, different on the east coast from the west coast, making the local Guide even more valuable.

Here are a few examples of common transition threats:

- A perfect resume will not get you an interview.
- You will (or may consider) alcohol to self-medicate through pain and to sleep at night.

- Your high school friends will not understand you and you won't fit in with them like you used to (unless some of them became veterans, too).
- The VA will not see you on a timely basis and will most likely treat you with a potent cocktail of medications that will affect you in ways you cannot predict.
- The kids in your college classes will irritate you to the point you want to drop out of school.
- You will not want to share the fact that you are struggling with anyone, and the sheer fact that you are struggling means you are failing, and you will never admit failure.
- Someone will ask you if you ever killed anyone.

Let's take that last one. If you don't prepare for the eventuality that someone will ask you this question, you won't be ready to respond. Whether you did or did not take the life of another human, you were part of a team that did. It is a very personal thing. This question might make you want to scream at the person asking the question. It might make you want to get in a fight. It might make you want another drink. It might send you into a funk that ruins the entire day, or evening. I find the best response is simply, "None of your business, and that's not an appropriate question to ask a veteran." Whatever your response, how much better to have thought of it ahead of time so you are ready when it comes — because you can't predict when it will come.

A transitioning veteran must understand that he or she risks becoming their own worst enemy. One in this position tends to sabotage oneself by asking why things are the way they are instead of what one can do to make things better.

Think back to my example of night land navigation. You stand there asking yourself, "Why can't I get back on azimuth?" and answers will come to you as excuses. "Because you're lost" or "your compass is broken" or "you're not smart enough to do this." Change that question to, "What do I need to do to get back on azimuth?" and the mind will work on creating solutions. If your answer starts with, "Because..." you're asking the wrong question and the answer will be an excuse to not make a positive change.

With every deviation from the transition azimuth, the actions required to get back on azimuth become harder, and the need to ask for help runs against everything the veteran has been taught. "Don't bring your problems to me — figure it out on your own..." "I don't want to hear any whining..." are commonly heard in the military. What is different about the military than civilian life is the system is set up so you always have a buddy, and in virtually every area (except for emotional fitness — there is still a stigma there) for you to ask for help from your buddy, is expected and encouraged.

The military system is also one set up rich in resources. There is a menu system for promotion, for assignment, and all one must do is follow the checklist, be tactically and technically proficient, be diligent, use initiative, and be persistent.

When a veteran tries this in the civilian world, one where there are few rules, resources are carefully husbanded and people seem less service oriented and more "me" oriented, the veteran struggles to grasp how to adapt to this change in conditions. Rather than ask for help and risk appearing incompetent, many continue flailing, burning resources of money and relationships until so much damage is done that recovering — getting back on azimuth — becomes a significant emotional and sometimes financial event for all concerned.

Rather than ask for help, the veteran builds a shell, a brick wall that seemingly provides protection from admitting failure, but one that drives the probability of failure much higher. Building this brick wall can lead to acute isolation — one of the greatest risks to a veteran's successful transition.

Isolation becomes a solitary confinement cell of one's own making. Isolation causes anything from diminished self-confidence and low self-esteem all the way to suicide. A veteran who is isolated is hidden from the community — trapped in a pocket of despair that goes on unseen, invisible to the community and seemingly cared about by no one. Commercials on TV, where celebrities proclaim their dedication to having our six, ring hollow. The veteran's nose is so close to that brick wall that there is no way to see there is space on the other side. Blue sky, green grass, opportunity and hope are completely blocked and seemingly out of reach.

Without hope, what's the use of living?

It may take years to get to this point, so is it any wonder that 65% of the twenty veterans who take their own lives everyday are over the age of fifty?

My own best friend Bill, who had proved so instrumental in helping me transition by being my mentor and staying with me even though I quit on him twice, had stopped answering emails and fallen out of contact with me in 2014. I knew he was going through a bitter divorce and that he had responsibilities heaped on him in his role as Chief Operating Officer of a construction equipment manufacturing company. Over the years we have fallen in and out of close contact, and always when we reconnect it's as if no time had passed in our relationship. This time, though, felt different and I allowed myself to be slightly put out that he hadn't been responding to me.

Then I got a phone call from a mutual friend asking when the last time was that I had spoken with Bill. "It's been awhile," I admitted. "Well, you should check on him," he said, adding that he was worried.

I called Bill, and he answered. I asked him how he was doing and there was a sorrow and heaviness in his voice that set off warnings in my head. I bluntly asked him if he had thought about hurting himself. There was a pause of about a half minute that told me everything I needed to know.

My God, I could have lost my best friend, and I'm supposed to be an expert at this.

What do you do when you find out that someone close to you—in this case as close to me as my wife—has thought about ending his own life? You tell him that you love him, that he is important to you, that you need him in your life. You tell him he can call you anytime, and if he needs you that you will be there. Then I asked him if he knew of the

Spartan Pledge. He did, because it's part of GallantFew. I reminded him what it said, and asked him to take it with me, and I had him watch the video of Boone Cutler telling of the genesis of the Spartan Pledge and how it works. You can see it yourself at www.spartanpledge.com.

The Spartan Pledge

"I will never take my own life by my own hand until I talk with my Battle Buddy first. My mission is to find a mission to help my Warfighter family."

It doesn't say I won't take my own life; that's the ultimate choice each of us has. It does say, however, that I won't do it until I talk with my buddy first. When a veteran takes this promise with a buddy they love and trust, one whom they would unhesitatingly shield from harm with their very own body, it becomes a promise that they will not break. It also contains a part "B," requiring the veteran to act. Action incurs purpose, with purpose comes hope.

Psychiatrists argue that such "pacts" are not effective and should not be relied upon, but they are not factoring in the veteran aspect of the relationship.
(http://www.psychiatrictimes.com/articles/no-suicide-contracts-suicide-prevention-strategy)

A Vietnam veteran who had served in a Ranger LRRP unit (Long-range Reconnaissance Patrol) walked up to me at an event at Fort Benning and blurt out that I had saved his buddy's life. I had never seen this Ranger before, so I asked him to tell me who he was and what happened. He went on to tell me that a Ranger he had served with in Vietnam, who he considered to be his best friend in the world had fallen into deep depression and was contemplating taking his own life. He made his buddy take the Spartan Pledge with him and they are both still alive, and his buddy's life is better now than it was when he was so low. It's a story I've heard repeatedly.

When life has piled up this much adversity, it's like standing with your nose to a brick wall. You can't see past it, and you can't see the green grass and blue sky on the other side. You also can't see that two feet to your right is a window, or to feet to your right the wall ends. Your battle buddy can nudge (or shove) you so you can peek around the wall, or through the window to see that in fact there is more than just a solid brick wall. Often the words, "I love you, brother…" open that window right in front of where your nose touches that brick.

Bill's nose was buried in that wall and it was hiding the joy that would yet come into his life.

A few weeks later, I made the drive across three states to spend a couple days with him. I gave him one half of a set of Spartan Pledge coins, a dog tag-size metal and enamel coin with the GallantFew logo on one side and the text of the Spartan Pledge on the other. It's a heavy piece, weighing more than a silver dollar. It has a hole just like a dog tag, so you can place it on your key chain. On the top edge is etched a small number, and each of the two coins in the set contains the same number. We both have coin numbered 6. It was good to see him, and we enjoyed our time together.

Fast forward a year. Bill and I are talking weekly, and his life has changed significantly for the better, although he was still dealing with a lot.

He has this great booming laugh that I love to hear. On this day, we're talking and I notice my half of the coin set sitting in front of me on my desk. I absent-mindedly pick it up and say to him, "Hey, I have my coin here on my desk." To which he answers, "I have mine on a chain around my neck." "Around your neck?" I point out that it's a heavy coin and ask, "Isn't that uncomfortable?" He answered, "Yes, it is, and every time I move it thumps me in the sternum and I remember."

While writing this, I texted and asked permission to use this story. He gave the okay, and closed his message with, "I love you, brother," and I responded the same. Virtually every time we talk on the phone, it ends with "love you, brother," or "love you, man."

The reverse of my business card contains the Spartan Pledge with a space to write in a name and a phone number. I point out the pledge to everyone to whom I give my card. It's powerful, it's real, and it works. And it helped keep my best friend alive.

Thank you, Boone Cutler, for your vulnerability in sharing why you started the pledge and your generosity in letting us share it, and for all the lives you've saved — the number which you'll never know.

I'm going to add one layer here. Scott Stetson published an article in 2017 about an experience he had where he became overwhelmed and had a "meltdown." Afterwards, he had a flash of inspiration and created a personal alert roster for his wife.

A veteran experiencing severe symptoms of Post-Traumatic Stress may become angry, agitated, and may act or appear threatening. Family members and friends are hurt, confused, and scared oftentimes not knowing how to handle the situation. Here's an excerpt, published with permission:

> *"That is when the clarity came. Not about what happened, but what I needed to do to set my family and friends up for success when it came to helping me if I ever am in that state again. We all went through establishing a will and a living will when we were in to take care of things if something happened to us but many of us have not taken the same care of providing something to our friends and family should a meltdown occur.*

"The number one thing I have advised my family is to not call the police unless I am actively going to harm someone or myself. This is not a dig on the police, it is a reality of our environment. Many law enforcement agencies are not prepared to deal with this type of event (I address this later in this chapter under the Functional Emotional Fitness section). *I have had close friends have family or friends call the police for help and they wound up dead. I don't blame the officers as they were reacting to a situation and in a couple cases individuals may have used the police to achieve a permanent and ultimately tragic outcome. I acknowledge that I had shut down. That means I am already not listening well and I could be perceived as resisting even though I am actively fighting full body contractions. Where does that lead if my hands are near or in my pockets? I don't want to put anyone in that position unnecessarily.*

"The next step is equipping my family with the contact information and sequence of contacting people that are "in the know" and can help. This is my new alert roster and it consists of local people I have spoken with in the past and facility contact information along with hours of operation. This requires significant footwork on my part, but it is the least I can do to ensure my family is prepared."
http://havokjournal.com/culture/life/military-preparedness/

It's time to stop being reactive.

*****

We can continue being reactive, issuing homeless housing vouchers, providing emergency financial aid, and enrolling veterans in in-patient substance abuse programs, or we can try something that will upend the transition experience — something "revolutionary."

GallantFew's "Revolutionary Veteran Support Network" proclaims and employs a different way. That veteran who is a transitional generation ahead can be a proactive Guide, providing preventive information in the form of a personal, local transitional AAR. His hindsight becomes your foresight.

How many societal problems could we prevent with this proactive approach? How many veterans will not be unemployed, will not become alcoholics, will not become divorced, but rather, will become leaders in their communities, have great jobs, great relationships, raise more stable families and pay a lot of taxes?

How do we create this proactive culture of veteran transition? We must have a lot of Guides, and the transitioning veterans must know where to look for them and connect with them.

The US Census says that in 2016, six percent of the population had served in the military, and that a half percent were veterans under 35. This country had been at war for over fifteen years by then, so it's logical to say that a half percent of the population are post-9/11 veterans. Most of them volunteered for military service knowing we were a nation at war, a war with no end in sight. Let's do the math. The Dallas/Fort Worth area had an estimated 7.2 million people in 2016. Six percent of 7.2 million is 432,000, a half percent of 7.2 million is 36,000.

That's 400,000 potential veteran Guides to a transitioning veteran population of 30,000. More than a ten to one ratio.

Sure, some of the veterans are World War II and Korean War veterans and they may be too old to help with professional networking—but these veterans could be invaluable in their ability to share their combat experiences. Not all the post-9/11 veterans need help—many do just fine as they move from their military to civilian lives, so let's agree that there are most likely ten potential veteran Guides for every one Future Guide. GallantFew uses the term Future Guide, because once a veteran works through transition he or she is now an ideal Guide for someone following in their footsteps.

That's a lot of resources. Resources that are local, connected, established; and in my experience these veterans are universally enthusiastic about Guiding a veteran just like themselves.

So, what does a Guide do?

Big Brothers Big Sisters has used one-on-one mentoring for decades, with proven outcomes of success. When they match a mentor to an at-risk youth, the kids are more likely to avoid drugs and alcohol, stay in school, and are less likely to endorse risky behaviors. The risk factors don't go away, but now the youth has a supportive, caring adult in his or her life.

http://www.bbbs.org/research/

These remarkable changes happen not because a "Big" forces the "Little" to behave a certain way; it's because the Big demonstrates a different way of life. The Big shares lessons learned and lets the Little know the Big cares. I can vouch for this by personal experience, both as a Little and as a Big (multiple times). I firmly believe this same principle works for veterans, and soon we hope to prove it with outcome measurements like Big Brothers Big Sisters.

In the veteran world, however, it is crucial that the relationship falls within certain parameters. It must be local—how can one mentor another by email or phone? You must be eyeball to eyeball. It must be based on commonality of service. I can have a challenging and productive conversation with an infantryman, a paratrooper, or a Ranger and I can be blunt about where I see their path leading. I believe the technical terminology is something like, "Pull your head out of your ass." I have the moral authority to challenge them, because I am one of them. I've proven myself in every field of endeavor they have, and I earn their respect as a result.

My conversation with a Marine is not as effective, because he can rightly tell me that I have no idea what he's been through, because I'm not a Marine. But if I can introduce a Marine into the equation—one who is local, with the same military training qualifications—then the dynamic shifts. The Guide simply asks questions, learns what the new veteran is going through, and applies life experience. A Guide is not there to "adopt." When the Guide uncovers a situation beyond simple coaching, GallantFew must be ready with professional resources such as alcohol treatment, PTS therapy, financial assistance, or any other diagnosed needs. GallantFew endeavors to develop relationships with many organizations so we can refer veterans to those organizations expert in the area of need.

If the VA would understand this, they could activate hundreds of thousands of veteran Guides in the Dallas area alone. Nearly every veteran I've approached has eagerly accepted the opportunity to help another veteran. Many want to hire a veteran just like themselves, because they know what they have accomplished with the benefit of their military background and to have another like them—that's good for their business! It's also a feather in their cap to introduce a younger version of themselves to a business partner or professional acquaintance—and when that veteran turns out to be a valuable member of the team, that makes them look really good.

There is a place for a non-veteran in this mix, too. Civilians possess tremendous business experience and skill, and most are genuinely grateful and appreciative for those who have served. They are priceless augments to the Guide, but without the moral authority and credibility of having the shared military background, they are less effective as primary Guides. In some cases, a non-veteran can worsen a veteran's situation by being too helpful—providing too much support, hesitant to be firm in challenging the veteran, and that can worsen the situation, making the veteran dependent on the benevolence of the other.

Multiple Layers of Community

Within the veteran community there are natural groupings according to service—the traditional service arms of the Army, Navy, Marine Corps, Air Force, and Coast Guard — but within each of these groupings there are many sub-groupings. Examples of these are units (the 25th Infantry Division, or the 9th Reconnaissance Wing); Military Occupational Specialties (MOS) such as 11B for infantryman or 88M, motor transport operator; Service Academy Graduates; even civilian education. When finding, and connecting a Guide with a Future Guide, the more layers of community the two have in common, the greater and more immediate the trust and bond becomes.

GallantFew has intentionally formed around sectors of service to help facilitate community structure and relationship building. GallantFew's Darby Project focuses on US Army Rangers — soldiers (and sometimes Airmen, SEALs, and Marines) who have completed US Army Ranger School and soldiers who have served in the 75th Ranger Regiment (or one of its lineage units — Vietnam LRRPS for example). The Raider Project focuses on Marine Corps special operations, re-designated Raiders in 2014 to reflect their WWII heritage and Marine Corps combat infantrymen.

Each "Project" is headed by a combat veteran who served in that sector and has the credibility — the moral authority — to build a population of veterans that helps itself. I have also discovered that what is effective for Rangers and what is effective for Raiders is different. I've learned to let each population lead itself, and the organic processes that develop are much more valuable than those forced on the population.

Karl Monger and Bill Cooper, February, 2017

Boone Cutler and Karl Monger, 2014

Chapter 12

Response-Ability

"You are responsible for two things every day:
Your attitude and your actions."
~ Grant McGarry

Imagine twin boys with an alcoholic father. One boy grew up to be an alcoholic. When asked why, he said, "I watched my father." The other never touched a drop. When asked why, he said, "I watched my father."

We all know people from the same family, raised in the same conditions who become very different adults. What is the difference? How can two twins look at something so differently?

I believe it's in the ability to choose. The ability to choose a response to any situation facing us.

I had the privilege of twice meeting the late Dr. Stephen Covey, author of *The Seven Habits of Highly Effective People*. Once, he signed his book for me and in the inscription, wrote "Be a trim-tab!"

Handing the book back to me he asked if I knew what a trim tab was and what it did. I had to admit that I did not.

He told me that on a sailing vessel, the trim-tab is one of the tiniest pieces of the ship, yet it is responsible for turning the direction of the ship by influencing a larger surface that does the heavy lifting. He put his finger to my chest. "Be a trim-tab."

The second time I met him he told me a story about perceptions. He liked to go to a diner once a week, enjoy the paper and a cup of coffee and look at his plan for the week.

This particular morning a young boy was running around the diner, knocking things over, making a lot of noise, and getting the attention of everyone there. It was impossible to try to relax and read the paper.

To make matters worse, the boy's father sat in a booth, staring off into space, oblivious to the havoc his son was creating.

Having lost his patience, Covey finally caught the man's attention and said, "Sir, can't you see your child is bothering people? Would you please do something to control your child?"

The father shook his head as if to clear cobwebs or perhaps to return to reality, and sadly looked at Covey. He apologized and said he had just come from the hospital where he learned that his wife — the boy's mother — would not recover from her illness and would soon die. He said he didn't know what to do.

If you're like me, you might have experienced the same emotional reaction to the story as I did. I wanted to duct tape the kid to the wall ... until I learned the reality of the situation.

Now I want to tell the father to go on thinking, make some calls, do whatever you need to do—I'll watch your son. Here, he can color on my newspaper!

What switch flipped in my mind (or was it in my heart)?

Outwardly, everything is the same. Diner, paper, coffee, brat, inattentive dad. But now *nothing* is the same.

What happened was I accessed our most powerful emotional state, that of Love. Learning to live in a state of Love is a bedrock of Functional Emotional Fitness which I will address in Chapter 13.

I believe each of us has the power to flip that switch and activate that most powerful emotional state on demand—the Ability to Respond, Response-Ability.

Chapter 13

Functional and Intentional Fitness

As noted in Chapter 11, I use the term "fitness" in the place of "wellness." Soldiers want to be more fit. Veterans want to be more fit. Fitness implies a level of competence, and fitness is something you can improve. Wellness implies a tolerable level of existence.

I heard an officer at USSOCOM's Warrior Care program say that PTS is an injury, not an illness. One can recover from an injury, but an illness can be terminal. I love this attitude, and it's one that resonates with veterans. I wrote earlier about the Army's eight battlefield operating systems, and I've developed five transitional operating systems, all identified as "Functional," means each applies to the way you live your life and do your job, and "Fitness" also means you can measure it and develop a plan to increase or improve that fitness level. It also implies that it doesn't just happen. Increasing a level of fitness requires hard, dedicated work.

When I hear "fitness," I think of getting faster, stronger, more capable—accelerating to the next stage of performance. Even now when I ride my bike, I want to maintain a faster average pace. When I rock climb, I want to climb harder routes.

Soldiers don't want to maintain, they want to accelerate to greater levels of performance. It's natural for them to think of this in terms of physical fitness, and the Army, especially the Ranger Regiment, has made great strides towards what it calls "Functional Fitness."

Now officially an "old guy" since I'm well over the age of fifty, I can say it. "Back in my day…"

Back in my day, our physical fitness test consisted of how many pushups one could do in two minutes, how many sit-ups one could do in two minutes, and how fast one could run two miles. Depending on the unit, they might also be measured on pullups and a timed road march with full equipment and weighted rucksack.

The test was scored on a graduated scale, based on gender and age. It didn't matter if you had to climb ladders or cook eggs, the test was the same for all.

The Ranger Regiment has developed the Ranger Physical Assessment Test (RPAT). By the way, anything the Ranger Regiment does must have the word "Ranger" in it (it's a thing). It is designed to evaluate and measure the Ranger's ability to perform physical tasks required of them in a combat situation, and is comprised of seven tests: a two-mile run; climb a twenty-foot rope; pull a field stretcher weighing 185 pounds for 100 yards; climb a twenty-foot caving ladder; sprint 200 yards; scale an eight-foot wall; and finally run one mile. This must be completed in under forty minutes wearing field uniform—boots, body armor, and a helmet. A soldier might hold the world record in pushups, sit-ups and two-mile run but if he can't complete the RPAT, he's not functionally fit.

It makes sense, doesn't it? Develop training specific to what you must accomplish; then measure your ability to perform to that standard. This makes it "Functional."

I mentioned above that I'm now officially an old guy. I'm not going to be taking the RPAT, as much as in my mind I'd like to think I'm capable of achieving that level of fitness— but when I look at what's important to me functionally, it's not pulling a stretcher and running miles while wearing combat gear. What is important to me is keeping a strong core so my back pain is reduced, and maintaining a strong heart so I can see my great-grandchildren. I've devised a diet and physical regimen to help me get there. What is important to you, functionally, right now?

If you're like most people you probably haven't thought about it. By not identifying it, and not creating an approach to becoming more fit, you are letting whatever happens to you happen by accident. Then ten years from now your waist size is ten percent greater, your arteries are ten percent harder, and you've shortened your potential life expectancy significantly.

Pure physical fitness isn't the only area we need to focus on, especially regarding transition. Whether you are a veteran or someone going through a divorce, job change, a death in the family, or another life-altering situation, transition is a supremely stressful time that most people simply try to survive. I want to challenge you to think greater than that, to start now before you are in a transition situation and put in place habits that will help you increase your fitness in each of the transitional operating systems.

Here are those operating systems again, in functional terms:

- Functional Emotional Fitness
- Functional Spiritual Fitness
- Functional Social Fitness
- Functional Professional Fitness
- Functional Physical Fitness

Is there redundancy in each area? Absolutely. For each to contribute to the greater whole, each is interconnected and relies on all the other systems.

Too many veterans' lives develop by accident. Let's change that by being intentional in becoming who we will be.

Functional Emotional Fitness

In *Paradise Lost,* 15th century poet John Milton said *"The mind is its own place, and in itself can make a heav'n of hell, a hell of heav'n."* The most powerful, lethal weapon in a warrior's arsenal is the mind. The Army's Ranger School pushes a Ranger student far past the point of physical exhaustion, where the mind screams at every step to stop, it's impossible, you can't go on—but you do. You learn to control (or at least mute) that voice telling you to quit.

I absolutely hated forced road marches. Loading up with a hundred pounds of weapon, helmet, rucksack, and other gear, and going for a twelve mile walk on asphalt while maintaining a brisk fifteen minutes per mile pace is sheer agony after a while. You start to develop hot spots on your feet, your gear digs into your back, you get dehydrated — but everyone pushes on. You must train your mind to overcome this physical discomfort, because that's all that it is, physical discomfort.

Everything a professional warrior does is geared at training the mind to overcome adversity to move further, faster, and fight harder — to survive, and win. In a sense this is easy, because you have a purpose, you have a mission, and you know if you fail that it's not just you that will pay the price — it's your buddy on your left and right. Failing them is not an option.

Then something happens that causes the conditions to change. A soldier gets a traumatic brain injury, or develops post-traumatic stress symptoms, or incurs a back injury. Perhaps a buddy is killed or maimed and the soldier's priorities change. Maybe he just wants to go to school because he's tired of war.

Now that most powerful weapon, the mind, starts to work against the soldier.

- "Why wasn't I stronger?"
- "Why wasn't it me that died, not him?"
- "They are going back to war; I'm a quitter?"
- "Why can't I sleep at night? Something's wrong with me."

225

But soldiers reach out to other soldiers and tell them they fear they are drinking too much, or ask if they also feel like a quitter, so they get compassion and support, right?

Not.

Perhaps it's the type of person attracted to the challenge, or the training, or a combination of the two, but an elite warrior does not admit weakness, does not ask for help. He or she is told repeatedly, "I don't want to hear you snivel" and to "solve your own problems." They are expected to be resourceful and resilient, and to be anything else is to admit failure, and failure is not an option. So they keep it inside until their life situation reaches crisis level. They've burned through financial resources, relationships, self-medicated, and only when they hit the very bottom will they even consider asking for help. Some of them never ask, a further reinforcement to the fact that more than half veteran suicides are old guys like me.

Often injuries, including Traumatic Brain Injury (TBI) and Post Traumatic Stress (PTS) symptoms are not reported because while it's hard to get into an elite unit, it's harder to stay there. Knees, back, joint pain, anxiety, difficulty sleeping, memory issues go unreported. This lack of documentation in their medical files bites them down the road because no evidence equals no VA disability.

These veterans are those in most need of VA services and are also the least likely to seek them, precisely because of this attitude—and when they do seek care it's because it's now a critical need, but it's not documented because it wasn't reported. It might be years before the condition is recognized and compensated, if the veteran fights through the process of appeals. This is a continual emotional beatdown.

Virtually every veteran I've spoken with reports this is the process they go through when they seek mental health care from the VA. First, wait for an available appointment, which is rarely within thirty days, regardless of the urgency relayed by the veteran (and when a veteran calls for mental health care, it's urgent). When the veteran meets with the care provider, it is most likely either an intern or a newly educated therapist. Few if any of them are veterans, and rarely does the veteran seeking care see the same therapist twice. The veteran leaves with multiple prescriptions for anxiety, depression, and sleep.

The VA employs prolonged exposure therapy, meaning the person experiencing distress is required to relay the intimate and often brutally graphic details of the combat event—every time the veteran meets with the therapist. Every time. The belief is that, like an old joke you've heard so often you no longer laugh at it, when the veteran has relayed the gory details so many times, he or she is simply so accustomed to it that it no longer carries the trauma.

To my uneducated mind that is the stupidest thing I've ever heard of. How can they possibly think that helps? How can you heal a wound if you constantly tear off the scab to look at it?

The VA says that sixty percent of the veterans who complete this therapy experience some improvement. However, more than fifty percent of the veterans drop out and do not complete the therapy. That would put the success rate much lower if they factored in those who don't come back.

The men I work with return home after a session such as this, drink themselves to oblivion and refuse to return.

If they do return, they get a brand-new therapist; they rarely see the same person twice.

If you saw a family member violently killed right in front of you, would you want to talk about that with a total stranger? Eight different strangers over eight weeks?

How close to malpractice or even criminal negligence would you classify treatment like this?

The stigma of getting help is not only real, it's validated by the system.

I've had conversations with battle-hardened Rangers, men of multiple combat deployments. Men who have met with and destroyed the enemies of our country and are of unquestioned bravery. Many come referred to me when they are having difficulty finding a job, or a relationship is on the rocks, or they are struggling with addiction. The last thing they want to do is tell me where and how they are failing.

When I talk with them, I describe several of my own personal experiences, and I describe experiences I've gained from others depending on the situation. I ask them if any of this sounds familiar.

"How did you know I'm going through that?" is a normal response.

"Because, it's normal," is my reply.

More than one Ranger has blurted, "So it's not just me," and burst into tears.

I tell them, "That's normal, too."

Then we can begin rebuilding their level of emotional fitness.

In 2015, GallantFew produced the award winning short film, *Prisoner of War*. In it, a veteran undergoes physical torture by an interrogator demanding he confess to his wartime sins. At the end, we discover the interrogator is the veteran, and he's been struggling in his mind, blaming himself for leading a convoy into an ambush. We are left wondering at the end whether he took his own life, and the film urges the viewer to reach out to a comrade.

The author of the screenplay, Ranger veteran Matt Saunders, sought not only to encourage veterans to reach out but to let them know they aren't alone in their thoughts, and he wanted to give others a sense of what a veteran might be going through. You can view this film here: https://youtu.be/BuvNOAutrBc

One of the priceless tools I've been given is Burris Emotional Fitness. Dr. Kelly Burris is author of nine books on behavior, and the developer of Subconscious Restructuring®. I was introduced to him by another Ranger veteran, Beau Chatham who has established Warrior Life Services. Beau had watched GallantFew's work through social media and thought that Burris would be a great addition to our resources, and he was right.

In a very brief nutshell, Burris Emotional Fitness measures a person's emotional fitness and coaches that person into a functional approach to improve their emotional fitness. It revolves around the questions we all ask ourselves, which usually are framed in the negative:

- "Why can't I lose weight?"
- "Why can't I stop drinking?"
- "Why did I ever leave the Army?"
- "Why can't I get a good job?"

If one can answer a question beginning with "because," then it's a negative question and the answer is going to be an excuse. That excuse gives you permission to stay right where you are. Reframing the question into a positive, empowered format such as "What do I need to do to get a good job?" requires the mind to produce solutions, and you can't answer that with a "because."

Recognizing, interrupting and restructuring that question into an empowered, positive question helps to activate our most powerful emotional state of Love. Not physical, romantic love but the love of the concept, "God is Love."

I prefer to boil things down into very easily understandable terms, so I'm going to give a simple example of how this technique worked for me.

I spent four days with twenty Marine Corps combat veterans at a retreat in Colorado in 2016. Over a two-day period, I took them through the Burris process and followed up with those desiring to go deeper into the techniques following the retreat.

The day after I flew home, I went out for a bike ride. I've learned since I can no longer run due to back pain and a prosthetic hip, that distance road biking is a great aerobic activity and burns lots of calories. I set a goal of maintaining a fifteen mile per hour pace.

Three miles into the ride, my ear buds announced I was at an average 14.2 mile per hour pace.

The first thought that popped into my mind? "Why can't I maintain a fifteen mile per hour pace?" Immediately excuses poured in:

- "You just got off a plane last night."
- "You're suffering jet lag."
- "You didn't sleep so well."

My subconscious was giving me permission not to achieve my goal.

I immediately caught it, interrupted the process and restructured the thought into an empowered form.

"What do I need to do to maintain a fifteen mile per hour pace?"

Solutions flooded in. "Remember to work the gears, put on faster music, hydrate," and by the end of the ride I'd gone more than twenty miles at a 15.6 mile per hour pace — the fastest I'd gone at that distance.

Burris Subconscious Restructuring © works, and learning how to access that most powerful emotional state of Love has helped veterans regain visitation with their kids, helped them identify better job opportunities, and helped them identify purpose which they thought had been lost.

Another example of services that I believe falls under the umbrella of Functional Emotional Fitness is the work that Dr. Carrie Elk does through the Elk Institute for Psychological Health & Performance based in Tampa, Florida. Dr. Elk uses a combination of the basic tools of psychotherapy in a non-traditional and very focused way to treat PTS. In just a handful of sessions, she helps veterans reprocess a fragmented memory of their traumatic event which is stored as sensory memory and which still generates a physiological/visceral response upon recall (veterans call this being triggered) into a complete memory (though still bad or sad) that can be recalled without being triggered. The result is elimination of PTS symptoms.

Dr. Elk's work is non-invasive, and does not include prescription meds for anxiety, depression, and sleep. I know Special Operations veterans who went through the VA system and became worse. She gave them the tools they needed not only to function, but to improve their level of functional emotional fitness.

Learn more at www.elkinstitute.us.

A key indicator of emotional distress is the inability to sleep. Sleep deprivation causes you to see things that aren't reality — and one of the greatest challenges for a veteran struggling with post-traumatic stress and traumatic brain injury is an inability to sleep. The longer this goes on, the harder it is for the veteran to process things clearly. This leads to alcohol abuse ("maybe it will make me sleep"), to arguments (the significant other doesn't understand why their loved one is acting this way), and the behavior can become more and more erratic. There is tremendous risk here, especially when the erratic behavior results in contact with law enforcement.

GallantFew does training for law enforcement officers on veterans' issues. What are they, why are they important to a police officer, and how do you prevent an interaction with a veteran from escalating and how do you de-escalate if a situation has gone sideways? Following one of the training sessions, I had a patrol officer email me. He had an interaction with a veteran, late at night, and his textbook method of interaction was to give commands and require compliance, using force if necessary. He decided to try one of our methods and, knowing the guy was a veteran, said simply, "Hey pal, how much sleep are you getting?" The veteran collapsed in tears and the result was empathy and transport to a VA facility when the usual would have been Taser and jail.

While writing this book, I worked with a veteran who is struggling with the fact that he's physically unable to do the things he's enjoyed in life. A broken back, bad vertebra in his neck, intense sciatica all combined to rob him of his identity.

As we spoke and he mourned for the loss of his former life. He said, "I sound like a wuss."

I pointed out to him that it was normal to feel the loss of his career much like the death of a friend. He sounded surprised then acknowledged that's exactly how it felt. I asked him what he was doing on a daily basis. He asked what I meant, and I repeated the question. He took it to mean what physical exercise did he do, and he told me that he tried to go to the YMCA but the physical pain made it difficult for him to even walk around the track.

I clarified. I asked what he was doing to help someone else. Was he volunteering or serving anywhere?

He paused; then said he wasn't.

I told him point-blank. "You keep going like you are, and you are going to die in ten years a bitter, broken man. You have to stop isolating, you have to get out, and you have to serve. I told him that helping an at-risk youth, or volunteering at a shelter, or even helping at a dog pound would put his life in perspective and help him cope with some of his physical and emotional pain. I also told him that what he was experiencing was common, that there wasn't anything wrong with him, he wasn't weird—and that he possessed the ability to make a change.

I could hear excitement in his voice as he realized he wasn't sentenced to the prison of his own home and there was opportunity for him to continue making a difference.

My final reminder to him: with purpose comes hope, and with hope comes reason to live. We hung up the phone with him already planning on what he was going to do to once again create that purpose.

A common theme to healing is the interaction with one caring person. Be that person. You'll be surprised at the healing that comes back to you.

Functional Spiritual Fitness

A Ranger veteran recently wrote about his transition experience on social media, and I reprint his comments here with permission:

> *"Transitioning out of the military is hard. It doesn't matter how connected you stay. You have no money, no mission, no relevance. These things are super hard to deal with. Some guys do it seamlessly, but the majority are lucky to not end up in jail or broke or both. There is not a day that goes by I don't wish I wouldn't have died in 03. I would've died a hero. Now, I'm a lonely nobody that nobody will remember. I mean, don't get me wrong, nobody would've remembered me from 03, but now I have to live to the best of my ability. It's no easy task. Men don't quit because their warrior spirit is gone. They quit because they've outlived their usefulness, and they know it. Some can hold on to their children's needs; some can't. The only thing to do is remember better days that sucked in an awesome way. Just sayin' Keep fightin' the good fight bro."*
> ~ *John Risley*

Boone Cutler, in the Spartan Pledge says, "A warfighter without a mission is a dead warfighter." Without a mission, one has no purpose, and without purpose how can one have hope?

Many veterans believe the most important accomplishment in their lives is now in their rearview mirror. Those days "sucked in an awesome way" because the sacrifice was for something worthwhile, something heroic — "to support and defend the Constitution of the United States against all enemies" and to be willing to lay down your life for a brother knowing he or she would unhesitatingly do the same for you. How do you transition to an environment where seemingly no one cares whether you even exist?

The VA's practice of declaring a veteran 100% disabled and unemployable reinforces this, because it requires nothing of a veteran in return. How can a person be officially declared "useless?" The VA does it every day. Unless a disabled veteran is bedridden and in a coma, he or she can be enabled to help someone else — and helping someone else gives purpose. We must help every veteran develop a sense of purpose or they have no hope.

In Viktor E. Frankl's classic book *Man's Search for Meaning* he explores his experience as a Jew in a WWII Nazi concentration camp. He noted that prisoners who should have been dead many times over somehow survived and prisoners who were among the healthiest didn't. He concluded that only one thing separated the two: hope. He quotes Nietzche: "He who has a Why to live for can bear almost any How."

How does one create hope where none exists? One must find a purpose. Cutler's Spartan Pledge contains the promise that "My mission is to find a mission to help my Warfighter family."

Don't have a mission? Your first mission is to find a mission.

Anything to create movement, planning, and thought will spark the will to grow — this feeds the spirit.

Functional Spiritual Fitness isn't about which faith you practice; it's about discovering why you are placed on this earth. What is your purpose, and how are you intentionally working to discover it? What are you doing to truly live?

I'm an amateur genealogist and have researched several lines of my family, going back to the 1400s. Do you know how many people it took to make you? If you go back two hundred years, that's your great-great grandparents and the total people it took to make you is thirty.

Go back another four hundred years to your 14th great grandparents: that's over 130,000 people.

Think for a moment what their lives must have been like. They went through tremendous trials and tribulations, those alive during the Plague saw half of their family and friends die horrible deaths. Your DNA possesses the genetic knowledge for survival. Acknowledging this and remembering those who went before you can be a powerful motivator to seek out your purpose, if you ask yourself deliberate positive empowering questions.

Prayer and meditation are also powerful tools that help one gain perspective and enable the search for purpose. I haven't gone into detail about my own personal faith on purpose, as I work hard to be a resource to all veterans, and I don't want my personal beliefs to get in the way. But I do believe in forgiveness and life after death, and there is nothing one can do to earn it because it's a gift. I also believe that our human understanding of this truth pales in comparison to what must be reality.

In the early 2000s, I met with the director of a faith-based organization that did global anti-poverty work. He bluntly asked me when the last time was that I worshipped, and as I thought to the last Sunday I was in church and began to answer, he interrupted me and said worship is a constant state of existence, of living in awe and wonder of creation and the force behind it. It's trying to live every moment being aware of and in touch with God.

Think back to the story I told about the kid in the diner and how your perspective changed immediately once you had additional information. The next time you feel overwhelmed, useless, without purpose, take a moment, breathe deeply, and look at the sky, the trees, the grass. Let everything fall into perspective. You can choose bitterness, regret and sorrow or you can choose joy and hope.

When you consider that you are a part of a creation bigger than yourself, your life can become more meaningful as you realize you are an integral part of all that surrounds you. You gain even more meaning when you reach out into your community and find a way to help another. People seek to understand God's plan for them and don't understand when life becomes difficult and painful. God is Love, and Love doesn't promise an easy and pain-free life; in fact, it thrives in the opposite. The Declaration of Independence doesn't promise happiness, merely the freedom to *pursue* happiness.

Functional Spiritual Fitness is a continuing, intentional process to pursue happiness by discovering one's purpose for being, and that purpose didn't end with departure from military service or any other life transition. Quick start your purpose by following Jesus' simple guidance: "Love one another."

Functional Social Fitness

> *"It was the one time in their lives that they were absolutely proud of what they were doing, and they were absolutely proud of their friends and what they were doing. It's a relationship of man to man ... and you go on, and the war is over, and you become the person you will be for the rest of your life. But inside of you, the time when you were men among men will never go away."*
> ~ Col. Paul Tibbets, pilot of the *Enola Gay*

I know firsthand how difficult it is to meet new friends and develop a social network. Sara took a job transfer to Dallas in late 2012. Besides the Army and a year in Albuquerque, I had lived my entire life in Wichita, Kansas. I office at home, because anywhere I can go online I can help veterans. To increase the degree of difficulty, I underwent total right hip replacement right before the move.

When I got to our new home, I knew very few people locally and absolutely zero in the immediate area. Because I work at home, I'm not going to meet anyone unless I initiate the effort.

I located my local Rotary Club, Metroport Rotary. Rotary is an international organization with a motto of "Service Before Self." I found all the information about when and where simply by googling my zip code and Rotary. Civic clubs such as Rotary let anyone visit and they'll probably buy your breakfast the first time you attend. It's an opening to meet new people, quickly. I found three veterans there, and quickly became friends with all three, as well as the other thirty club members. Some of these friendships have helped build the local veteran community, and now we're active in the Metroport Veteran Association, a free informal veteran networking group that meets the 2nd Saturday of each month at Meat U Anywhere BBQ in Trophy Club, Texas. The club also supports local impact grants, and asked members to submit grants. Because of this program, Metroport Rotary has supported GallantFew's veteran indoor rock climbing program at a local climbing gym.

One of the veterans benefitting from the climbing program is Nate. Nate gave permission to share this:

> *"I know veteran groups exist for everything and so I searched for one that does rock climbing and this is how I found Karl. It really took everything I had to enter that building, there were so many people… When I got geared up and linked up with Karl on the wall I was almost ready to run for it but then it was my turn. I really don't know how to put into words the feeling I got once I started climbing,*

but it took every bit of me to climb, for once I could think exclusively about what I was doing, I felt in control, like I was playing chess with myself. Almost as if the two parts of my fighting mind are doing separate things, like one is working on moving and keeping my body on the wall while the other is working on the next hold and what it needs to tell the other part to do to get to it. I felt peace in my head like it was taking a break from the fight for once. When I came down I started crying because even for a minute that silence in my head meant so much to have finally found it. The gym is still a pretty terrifying place for me but that peace is worth fighting through it for."

Then I gave the belay end of the rope to Nate and I hooked in to the carabiner. I wanted him to see that I trusted him with my life while I climbed. Many veterans don't get to experience being trusted with something as precious as another life. It takes them back to the buddy system. I got you, you got me.

After only a few months Nate became one of our strongest climbers, eager to jump in and coach another veteran on the finer points of technique. One more quote from Nate:

"This is a game changer for me."

If I hadn't made the conscious decision to go make friends in a very new place, I wouldn't have joined Metroport Rotary, they wouldn't have sponsored the climbing program, and Nate would still be searching.

Alcohol

The late Pulitzer Prize-winning journalist, Jimmy Breslin once said, *"Whiskey betrays you when you need it most. You think it will fortify you. But it weakens you."*

Most new soldiers enlist between the ages of seventeen and twenty-two. These young men and women are highly impressionable and without doubt still maturing, and the social skills developed there may or may not fit in well in the civilian world.

Soldiers learn how to drink when they get to their units. This education doesn't include which wine goes with which cheese or the finer points of a microbrew. They train hard, they live hard, and they party hard. There is tremendous peer pressure (and sometimes sergeant pressure) to participate. While the military's top leadership levels discourage drinking, it remains a staple of a young soldier's existence. This can have a devastating effect, unlocking alcoholism and associated misbehaviors like poor work performance, self-medication or even a DUI.

When that soldier leaves active duty, he or she carries learned alcohol behavior with them. I've worked with veterans who drink as much as a handle of hard liquor a day, every day. A handle contains the same alcohol as 40 shots—in a day! Someone consuming that much alcohol must undergo detox; if they simply quit their body will go into shock and they could die.

How does one afford a handle a day? 100% VA disability with no expectations. No purpose, no hope.

A critical function of a Guide is to demonstrate that alcohol does not need to be the centerpiece of social interaction, that there is responsible use of alcohol, and to encourage the veteran to seek help if necessary. In the GallantFew network, the Guide alerts us and we get to work identifying resources and encouraging the veteran to seek appropriate care.

Dysfunctional Is Cool

There are many popular social media sites and apparel companies that trumpet "dysfunctional" as a desirable label. Our own veteran community sends a message to non-veterans that veterans are unstable and possibly dangerous individuals.

In March 2017, every news service reported on a scandal that had hit the Marine Corps. Active duty and veteran Marines were sharing and commenting on nude pictures of other Marines (usually female). There will very likely be charges and lawsuits filed as the extent of this grows. It's easy to make poor decisions in cyber social media; it's easy to go along with a crowd.

It's also easy to misinterpret a text, email, or social media message. Too often we read our own emotion into another's words. I've embarrassed myself by reading a message one way, when the sender intended a meaning opposite my interpretation.

Many active duty soldiers don't have social media accounts due to security concerns. Stepping into cyberspace becomes another trip into night land navigation. Here, a Guide helps establish and maintain azimuth.

You're Not the Cop (Unless You Are a Cop)

Transitioning from a strictly defined environment to one not as defined or structured differently, is difficult. A strictly defined environment has rules that initially make no sense but later are revealed to be important to the overall functioning of the greater team. In the civilian world, many veterans see people breaking the rules as people openly displaying disrespect. When I see someone litter, I want to go make them pick it up. When I see someone cut the line, I want to put them in their place.

I know a veteran who was intent on paying attention in a college classroom, and the other students were being disruptive, noisy, disrespecting the veteran's desire to learn. Having had enough, he slammed his fist down on the desk, shouted at the top of his voice for everyone to shut up (there were most likely some curse words interjected) and pay attention to the professor. Who do you suppose got in trouble? You probably guessed it, he did. Got suspended from school, now he's labeled as a dangerous, unbalanced veteran prone to violent outbursts.

Is it any wonder that half of the veterans drop out of college, despite what I believe are the best GI Bill educational benefits afforded veterans in history?

A Guide can ask questions about school and engage in an open discussion that may help the veteran put things into perspective and keep an eye on the long-term, which is completing studies and getting a great job.

A veteran will learn that the rules out in the civilian world are different. You're not the cop (unless you are a cop), and telling other people what to do will only invite conflict and reinforce an unstable veteran stereotype. It's much better to be warned and prepared and perhaps avoid unnecessary conflict.

In 2016 Chris and I had just finished a climbing session and we went to our favorite "taco Tuesday" joint, which featured tacos for $2.50. As we sat at the table, in walks a young man with a girl, and he has the American flag dragging from one hand. He sat down at a table with a half dozen other men and one other girl, and tied the flag around his neck.

Chris gave me a look, and I got up, walked over to him and asked very calmly and respectfully for him to treat the flag with respect as it had decorated the coffins of a number of my buddies (I was wearing a ball cap with a Ranger scroll on it). He looked at me and said he was wearing the flag for a soccer match getting ready to air, and he was supporting the USA.

I told him great, but to untie the flag, and if he tied the flag around his neck again I would take it away from him.

I could feel the tension of the other guys bristling. I told him, "I don't want to fight. Just treat the flag with respect," and I turned away. He made a noise, and I turned back toward him, put my finger in his chest and gave one more warning. "Tie it around your neck, and I will take it away."

He didn't tie it around his neck, we finished our tacos, and left.

At the time I was 55 (although my wife tells me I look much younger). Not the smartest thing to do, taking on a half dozen men half my age, and it's proof to me that I still need to work on my social fitness. I did, however, avoid a fight and I hope I educated someone on what not to do to the flag. It's nearly impossible for me to walk past a wrong without trying to correct it.

Functional Professional Fitness

The military creates and enforces structures around soldiers' lives within a clearly delineated process for success. Attend certain schools, earn certain awards, accumulate enough promotion points, and you are offered the opportunity to appear in front of a promotion board. Perform well in front of the board, and you will be promoted. There is a process and a timeline for every personnel action, and with promotion, comes opportunity. A common saying at a promotion ceremony is that one is promoted not based on past performance, but on demonstrated future potential.

Any active duty person can map out his or her entire professional twenty-year career before it even begins, and barring unforeseen events, could follow that plan to a successful height, retiring as a Lt. Col. or Master Sgt. or Sgt. Maj., or the equivalent in their branch of service.

It should come as no surprise that a veteran expects to leave the military and move, at a minimum, laterally into civilian employment. A sergeant in a team leader job in an infantry unit is the direct supervisor for four soldiers, and the sergeant might be twenty-two years old. That sergeant is probably not going to walk into a civilian job with that level of responsibility and authority.

Few if any civilian jobs are awarded based on demonstrated future potential — unless there is a military veteran in a leadership role in that company. That veteran leader will understand the dynamic and place a greater value on the demonstrated characteristics of military service, and might be more inclined to take a chance on a veteran. Just like Bill did with me.

Military transition training places a premium on creating the perfect resume. A great resume will get you a great job, right?

What gets you a great job is a robust network, and a veteran leaving active duty and returning home no longer fits in like he or she did. If they serve one or two enlistments, then when they return home their friends are just graduating college, or working a blue-collar job. Neither are ideal networking resources.

Here again the Guide plays a crucial role. Remember, the Guide has completed school, landed a job, and is established in the community. He or she belongs to civic organizations, the Chamber of Commerce, business networking groups, and even athletic leagues. The Guide can turn-key a professional network for the transitioning veteran, simply by making introductions.

Steven Barber (see A Tale of Two Rangers) openly admits transition was much harder than he anticipated, and that he didn't fit back in to the old neighborhood as he expected he would. The GallantFew network, and I as his Guide, opened the doorway for connections. With connections came opportunity. Now Steven, a former specialist with responsibility for four soldiers owns his own business and employs five people less than four years after leaving the military.

Education

Education is an important component of Functional Professional Fitness. An artillery veteran going to school to study law should be guided by a local artillery veteran who is now an attorney. A Marine Corps veteran studying business should be guided by a local Marine veteran who owns a business. Whether that transitioning veteran wants to be a teacher, a plumber, or a candlestick maker, a similar Guide can be priceless in helping achieve success.

Besides reinforcing the value of staying in school, the Guide contains the keys that also open connections to industry professional associations, scholarships, summer jobs, internships, and potential post-graduate employment.

Service-Disabled Veteran-Owned Small Business

SDVOSB for short (the only acronym in the government that isn't pronounceable), Service-Disabled Veteran-Owned

Small Business is a program that President Bush (43) signed into Public Law in 2003. It sets aside three percent of federal discretionary spending for award to SDVOSBs. The program doesn't care the color of your skin or your gender. It cares that you served, and you got hurt doing it.

Few veterans are aware of the program, and rarely is it covered during transition training. A zero percent disability rating counts, meaning the injury is acknowledged by the VA but causes no pain or limitations. The veteran doesn't have to be receiving disability compensation.

The service-disabled veteran (or combination of veterans) must be in undisputed control of the business, with a minimum 51% ownership.

There are three levels to SDVOSB. The premier level is being accepted by the VA as a "verified" SDVOSB and placed on a list maintained by the VA and is a requirement for the SDVOSB to compete in preferred status (meaning sole source, qualified for non-competitive awards up to certain levels). This is a strict process whereby the VA confirms the veteran does have a recognized disability rating, that he or she is 51% or greater owner and has control of the business, and further, that he or she is qualified to operate that business.

In 2016 I began the process of obtaining the VA's certification as a Service Disabled Veteran Owned Small Business. In April, 2017 I received notice that I had been approved, and my small business Transitur, LLC is now listed on www.vip.vetbiz.gov along with the blue wreath logo of authenticity. The VA has greatly improved the process since I first went through it in 2010, but still required me to justify how I could run my small business (I'm the only employee—and so far an unpaid one at that) while I held a full-time job. I was required to justify why I wasn't spending forty hours a week.

As owner of the small business, it should have been none of the VA's business whether I worked one or a hundred hours in *my* business, and so I told the case manager assigned to me. She didn't disagree with me, but I still had to provide a memorandum justifying my time.

Why is the VA making judgements regarding the capability of a veteran to run a business they claim to own? That falls under the realm of the SBA (Small Business Administration). The SDVOSB program should be moved to the SBA, and the only role the VA should have is verifying the person is a veteran with a disability rating.

Unfortunately, there have been too many cases of fraud, of non-veterans propping up disabled veterans as business owners, and using that preferred status to illegally line their own pockets. One last comment: my impression of the process of obtaining certification, while a pain, was that the VA was helping me provide the documentation and evidence so that I could achieve the status, and not placing obstacles in my way to prevent that.

The second level to SDVOSB is the rest of the federal government that is not the VA. The law applies to the entire government, but for some reason only in working with the VA is actual verification required. A veteran can simply self-designate on the government's purchasing management website, www.sam.gov (SAM stands for System of Acquisition Management), and look for business. Most federal purchasers, however, look for the VA seal of approval as assurance they are working with a legitimate SDVOSB.

The third level is comprised of companies that hold large federal contracts. Baked into their contract are requirements to sub-contract with federal designated under-utilized businesses and SDVOSB is one.

I encourage and help veterans to establish their SDVOSBs first at the third level which has the potential for immediate business while working the laborious process to achieve the first level. It's a process involving filing with the local Secretary of State, obtaining an EIN, then a DUNS number, then a CAGE code, then applying for VA verification. I'm not going into detail on these. If you want to know more, visit www.gallantfew.org, and we'll walk you through the process (for free).

Need a good format for a business plan? The same operations order format that formed the basis of planning for the D-Day invasion of Normandy works great for business planning, too. This is a very brief outline, just for illustration purposes.

> Paragraph 1:
> Situation, enemy and friendly. This is your competition, suppliers, and other environmental factors that might affect the business.

Paragraph 2:
Mission: Who, What, Why, Where, and When (but not How)

Paragraph 3:
Concept of the Operation. This is the How — the meat of the plan. Detail how you will provide your service; think through all stages of the business from start-up to where you would like it to be in five or ten years. List your fire support — friendly businesses or assets that will help you out. List your emergency plan and how it kicks in.

Paragraph 4: Service and Support. This is your logistics section. Products, budget, finances, all goes here.

Paragraph 5: Command and Signal. Identify key personnel, their roles and reporting structure. Determine how they communicate, to include social media and web presence.

Once written, go back through and remove all the military headers and you'll be surprised how well organized it will appear.

Financial Management

Managing your money is an important skill that is taught differently in the military. When I was on active duty, budget management equaled "use it or lose it."

Failure to spend all funds allocated could mean loss of those funds next year, so it would be best to use them all and ask for anything else available. This always sparked a fiscal year-end spending frenzy to make sure all available funds were committed. If you didn't spend all the money you were allotted for the year and asked for more, you stood to receive a reprimand.

If you manage your home budget the way the military manages its budget, you're bound for disaster. When you're on active duty, you have a predictable paycheck deposited on the 1st day of every month no matter what, for as long as you remain in the service which subject to extreme drawdowns is determined by your enlistment contract.

Civilian employment doesn't work like that; there are few if any guarantees, but most veterans move into that next job with the mindset as if it's as stable as a military enlistment and they fail to start saving for a potential unemployment period.

I've worked with veterans who had such difficulty getting hired, that they enrolled in school just to receive the GI Bill living stipend. They took easy classes and got some cash, but did nothing to further their education. If they blew off their classes and flunked or got an incomplete, the VA reached into their bank account and recouped the entire tuition paid in one fell swoop and often without warning.

Some veterans win years-long battles with the VA, finally obtaining a service-connected disability rating. The VA back pays to the date the claim was originally filed, which can turn into a substantial tax-free check.

A Guide has worked through factors like school, job, and profession, and has personal experience with the VA system or knows someone that does. This knowledge can prevent many issues that will crop up later.

Reading

A critical component of professional development in the military is studying one's profession. Every commander has a "recommended reading list" that he or she requires subordinates to read.

I was tempted to publish my recommended reading list at the end of this book, but I realized as I compiled the list that many of the works I would recommend didn't exist a year or two before publication. This is a dynamic space and there are many excellent works pertaining to veterans and to the general population that are highly valuable to not only veterans but also and anyone seeking to better understand a veteran or a veteran's perspective.

With that in mind, I'm publishing a recommended reading list on karlpmonger.com, and will expand that will reviews and blog posts about especially important sections. If you know of a great work that I've omitted, please message me so that I can consider including it on my list.

Functional Physical Fitness

I've given numerous examples already of Functional Physical Fitness because it plays a key role in soldier's daily routine and it's easy to identify with. It's also one of the first routines to drop by the wayside once on the veteran side of service. Keep this in mind:

"Paper stacked a sheet at a time will eventually reach the sky."
~ Unknown

A daily routine of exercise, controlled diet, and responsible alcohol consumption is the daily routine equivalent of stacking sheets of paper. You don't see the daily progress but decades down the road it can be the difference between a healthy heart and a heart attack. It's the ability to handle stress without your blood pressure endangering your life. It's being able to walk your daughter down the aisle, or play ball with your grandson. It's also a life of reduced pain, for the back injuries so prevalent in soldiers and Marines become more painful as the gut expands and core softens.

It's about being intentional. It's about being functional.

Why are you trying to bench press and squat the same weight you did when you were humping ruck sacks and machine guns up and down mountainsides?

Look at the activities you enjoy and devise a training plan to increase your level of fitness for that activity. I love to run; once I was fast and had endurance. Then my back pain caught up to me and one day I learned I had no cartilage remaining in my right hip. Doc says I'm not to do high impact activities (like running and racquetball, which I also love).

I enjoy a great craft beer occasionally, and I like pizza. If I don't figure out a way to get some strength and cardio going, I'm going to become overweight, stress my heart, unbalance my spine, increase my pain and add misery to my life.

I discovered two things that work for me—indoor rock climbing and distance road biking. In the three years since I began climbing I've reduced my back pain by strengthening my core and improving my flexibility, and I've dropped some pounds as I've added miles on the bike. I found some other veterans that enjoy each activity, and now every week I meet them and we climb and ride together. This has deepened friendships and brought me new ones like Nate.

If you don't make it fun, you won't do it. If you don't make it functional, you'll get injured. Make it a routine and you might live longer, enjoy life more and feel better physically and emotionally.

If you have trouble figuring out what works for you, contact us at GallantFew. We have volunteers who are subject matter experts in all areas, especially Functional Physical Fitness. We'll help you create a plan that works for you.

Chapter 14

What If?

What if we realized the military should focus on military tasks, not on trying to make soldiers into civilians?

What if we realized that a centralized one-size fits all transition program is less effective than a local effort?

What if every person leaving active duty was given the opportunity to connect with a hometown veteran Guide, just like himself or herself, before he or she ever left their final duty station?

What if this Guide had received an orientation on mentoring and that orientation included a list of available resources, making him or her well prepared for the initial contact?

What if the community, in partnership with the VA, held monthly or quarterly veteran welcome home events where the residents and businesses could show their gratitude by immediately including and connecting the new veteran?

What if local and national corporations encouraged their military veteran employees to guide local transitioning veterans? A large corporation with 10,000 employees across the country will have hundreds of already transitioned veterans. These veterans are ideally situated to connect with and guide veterans just like them, following the same path.

What if colleges and universities developed their networks of veterans who are graduates and connected them with undergrad veterans, just like themselves?

What if veterans received access to mental and physical health care in the same timely standards expected and received by a person with private health insurance?

What if veterans so severely wounded or injured that they are declared 100% disabled were treated and were expected to continue to act like contributing members of society, and were given a role to play within the limits of their disability?

*****

Throughout this book I've given examples of transition issues I have experienced personally and that I've encountered helping other veterans. We repeat the same mistakes individually, and we repeat the same mistakes nationally. Veteran unemployment, homelessness, and suicide remain major issues. It's time to stop doing what doesn't work.

Veterans return home from active duty and have difficulty fitting in, building a network, and locating and accessing resources. Let's create a system where veterans are welcomed, connected, and included.

I am inviting you to help.

Transition is local, yet we act like it's long distance. A Killeen Texas Chamber of Commerce 2016 study found that 74% of veterans leaving the service did not plan on staying in the area of their departure station. These veterans go all over the country, many back to where they grew up. The tools they are given in the heart of Texas might or might not apply in Los Angeles, Chicago, New York, or Miami — but all those locations have transition programs pushing veterans to other locations. Let's start by having soldiers go through transition where they intend to live, not from where they depart.

Why should a local employer support a job fair in Killeen, knowing that most of the attendees aren't planning on staying in the area? That employer wants to find the people that are coming back to the area, and we don't have an established system to facilitate those connections

With a local transition focus, the military, the VA, local governments, businesses, and Chambers of Commerce could collaborate to develop welcome home events. Imagine an ice cream social in a town square, business people wanting to hire veterans could wear blue name tags, veterans seeking a job could wear green name tags. Green tags, go meet some blue tags!

I call on the VA to establish a core of volunteer mentors in every community, train them and make them available to meet with and guide veterans as they transition home.

I call on the VA to partner with local Chambers of Commerce and local governments to create events where veterans and the community can connect.

*****

Some veterans have a great experience with the VA. They have rapid access, are pleased with their team doctor, and wouldn't change a thing. In my experience, however, this is not the norm.

Veterans face interminable waits for VA health care. In the Dallas/Fort Worth metro area of over seven million people, spread out over more than 9,200 square miles, there are three major facilities: Dallas VA Medical Center, Sam Rayburn Memorial Veterans Center (Bonham Texas, northeast of the metro area and a Fort Worth Clinic. There is a small clinic in Denton, Texas. The Dallas VA Medical Center is the only full-service facility.

I'll use my own personal experience as an example. The Denton Community Based Outpatient Clinic (CBOC) is the nearest to me, and it's located over twenty miles away. I go there for my annual physical to ensure my enrollment in the VA system stays current. Several days before my physical, I must go in and submit lab tests. The last time I did labs at the Fort Worth VA Clinic there were over fifty veterans waiting in line ahead of me. To get to the nearby Denton CBOC, in rush hour traffic it's a 45 minute drive one-way, and I'll spend one to two hours at the clinic. That's two half-days I must take off from work. Would you be pleased if that was the way your private health care insurance worked?

Contrast that with getting a physical through my wife's company's health insurance policy. I live fifteen minutes from the family practice clinic (in traffic) and I'll spend thirty minutes there. That's an hour each day, total of two hours versus two half-days.

If I have a sinus infection (and I'm prone to them because of the asthma I developed on active duty), there is a drop-in clinic at each facility. With no appointment, that's a full-day process. The last time I had a sinus infection I used my wife's health insurance, called my nearby doctor and was in and out in less than thirty minutes.

The last time the VA reviewed my physical condition to re-evaluate my disability rating I spent an entire day at the Fort Worth Clinic. A doctor spent an hour with me, 45 minutes of which were small talk about how to research family history before he decided it was time to check me out. He never got out of his chair, never measured my degrees of flexion in my spine, mumbled something about taking care of my leg (which I hadn't mentioned) and sent me out. I wanted to talk about my hip replacement but that wasn't on his agenda. I was so mad when I left, I didn't want to talk to anyone else from the VA. All that wasted time and he didn't even listen to me.

The result was a reduction in my rating for my back injury (which is most definitely not better) and an increase for sciatica in one leg, which we never discussed.

Unfortunately this story isn't just my story. Veterans wait hours and hours at the VA for care, and they get frustrated.

They voice that frustration and it seems the wait becomes even longer.

Over the years my levels of pain have increased. After six months of physical therapy at a Regional VA hospital, the director of the physical therapy clinic decided I wasn't going to improve, so further physical therapy was denied. I was given liberal doses of hydrocodone, a powerful opioid. As my pain levels increased, the VA simply doubled the dose. I became a frequent visitor to the pain management clinic, which could not do anything to help, except prescribe more drugs. The hydrocodone was affecting my work, my thoughts, my mood, my memory. I asked for an alternative that wasn't an opioid, and finally was switched to Tramadol.

When I moved to Texas I used my wife's private health insurance to seek care through the private medical system. Most every veteran I know who has a private insurance option uses it, and only goes to the VA just frequently enough to stay in the system. In 2014 and again in 2016, I paid out of pocket thousands of dollars in co-pays to have a series of radio frequency ablations done on my lumbar and cervical spine through a Texas spine center. If I tried to get it done through the VA, I'd still be waiting for the appointment. These treatments, although temporary, reduce the pain level enough to allow me to participate in the activities I enjoy, and I only take the Tramadol now when I have a severe flare up of pain.

I call on the VA to allow veterans to receive the community's best care—mental and physical—managed by a local family practice physician. The VA can preauthorize treatment for service-connected conditions just as my private insurance preauthorized the spinal procedures ordered by my doctor.

What if we decided to make *top priority* those who have put their civilian lives on hold to serve our country, and sacrificed so much to fulfill the commitments they voluntarily made?

What if we decided to use common sense?

Chapter 15

Let's Use Some Common Sense

What are we going to do about it?

Let's start by recognizing veteran transition is a local responsibility, borne by the VA, the community, and the veteran.

On a national level, reverse the process. Provide veterans the universally applicable transition information related to the military aspect of benefits at departure station, and require the veteran to report in to a local transition center wherever they go to live for final training and connection to a Guide. Hold a portion of the veteran's final pay until this is complete, to ensure it happens.

Aggressively develop local veteran networks. The VA must take the lead in this because the VA knows who the veterans are and how to contact them. Engage local media — radio, TV, newspapers — to put out a call far and wide, and to acquire veteran Guide volunteers. Train them and connect them.

Local governments: identify veterans in your community and encourage these veterans to meet monthly in a no-cost, public, no pressure forum.

Once-a-month Saturday breakfasts are a great way to do this, and as the word spreads, more and more veterans will come. GallantFew can help you organize and grow this activity. Recognize the reality that the problems accompanying failed veteran transition will cost your community money. An unemployed veteran pays fewer taxes, buys fewer goods, his property value will decline as he undergoes foreclosure or eviction, a divorced veteran's problems compound, single parent kids get in more trouble, and a veteran self-medicating may kill someone on the road.

Corporations: encourage your military veteran employees to make a difference in their local communities by guiding a transitioning veteran. What a great way to be a valuable community service and enjoy good public relations for the company.

Colleges and universities: identify your graduates and undergraduates who are veterans. Hold open houses and invite your alum veterans to meet the undergrad veterans. Match them up by branch of the service and discipline of study. Your alum veterans are ideal guides and tutors for the undergrad veterans, and your alumni will be more closely connected to and involved with your institution.

Veterans: assume none of my recommendations are going to happen. Don't wait for someone to tell you to do something. Do what you've been trained to do—lead! Start organizing. Look for veterans on your city councils, leading local businesses. Ask them to be advocates for monthly get-togethers. If they won't do it, start one yourself. Find a local breakfast spot, coordinate with the manager, leave flyers and post social media events. Reach out to other veterans you know and encourage them to bring veterans they know. If you are a college graduate, contact your school and help them set up a program matching alum to undergrads. Sign up as a GallantFew (or perhaps in the future a VA) Guide.

Stop glorifying the image of a dysfunctional veteran. Start taking care of your veteran community in a positive, empowering way.

It only takes one to start. One person can make a tremendous difference.

Cory Smith

In 2012, a US Army Ranger, Corporal Cory Smith, faced the reality that he had to leave active duty. Going through a separation from his wife and infant daughter, he had to choose between being a father and being a Ranger. He decided to end his time in the military and focus on being a dad. Torn between these two worlds, Cory would come home from the Army at night and go out running for miles and miles, trying to make sense of things.

Alone in an empty apartment, Cory mourned for the end of his military career and dreaded the separation from the men he knew and loved. As he slept, Cory dreamed that he ran and as he ran he tried to tell people how hard it was to go from being a soldier to being a civilian. He knew from some of his friends who had previously transitioned that it was going to be much more difficult than he'd been told by the transition training. They were struggling, and it made him afraid of the unknown future.

Cory shared his dream with his pastor, and his pastor suggested that God was giving him a mission, to tell others about his struggle. Cory decided to run home.

Home was 565 miles away, and his journey would begin in January. Cory ran twenty miles a day until he developed a stress fracture, then he walked until he got a bike, then he biked, and he completed the entire journey in a month, accomplishing his goal. He partnered with GallantFew to provide support and visibility and appeared twice on CNN's Robin Meade's "Salute the Troops." His Congressman invited him to the State of the Union Address. The Mayor of Indianapolis declared February 8th "GallantFew Day." He reunited with his daughter in an emotional ceremony and raised tremendous awareness for veterans' issues.

Cory Smith and Karl Monger on the last day of Cory's run,
February 28, 2012

The following year GallantFew created "Run Ranger Run," an event for teams of ten people to form a team and try and recreate Cory's 565 miles as a group, walking, biking, running. Amazing and touching stories and wonderful relationships have grown from Run Ranger Run, and in 2017 over 1,300 people nationwide participated. Over a half million dollars has been raised since 2012 and have allowed GallantFew to grow and reach thousands of veterans.

After the first official Run Ranger Run event, it was time for another Ranger Rendezvous, an event that happens every two years when the Regiment changes command and Rangers converge on Fort Benning. I invited Cory to return to Fort Benning, to tell his buddies still serving that when they left active duty we would be there to help them.

Cory agreed to come and I took him out to dinner the night before the main family barbeque where he would see his old friends. He didn't eat much, and didn't talk much. The next morning I picked him up and as we drove toward the event, I could tell something was wrong. Cory looked at me and said that he didn't feel well and perhaps shouldn't go.

I immediately understood, because I had felt that way when I returned to Ranger Rendezvous as a guest of then Col. Votel. How would I, a quitter, be received? Now here was Cory going through very similar emotions as I had.

I asked him if he was worried how he'd be received, and he confirmed my suspicions, saying that he was the guy who had left the team, whose wife had demanded he quit the

Army, and these men went back to war while he stayed home safe and sound.

Returning and connecting with my old comrades was an important part of my personal healing, and laid the foundation for my future work. I intuitively knew it was just as important to Cory, and I asked him to go, to walk in the door, and if at any time he was uncomfortable all he had to do was look at me and we'd leave immediately. Cory agreed.

We parked near Lawson Army Airfield and walked into Freedom Hall, a hanger large enough for a thousand Rangers and their families to enjoy barbeque and music. As we walked in the door I heard a voice from across the hanger bellow "Cory!" A grin broke across Cory's face as he strode across the space. He was enveloped in bear hugs from his buddies. I didn't see him again until the event was ending and it was time to leave.

Cory was experiencing the self-induced separation that is so common among these elite soldiers. A retired Ranger 1st Sgt. once told me that the only honorable way to leave Regiment was in a box with a flag on top. Remember this from Chapter One:

> *"That Ranger now looks back on his Ranger time with embarrassment rather than pride and avoids the Ranger veteran community because he doesn't want to confront that shame, that failure.*
>
> *"It's compounded if a Ranger buddy subsequently dies or is seriously wounded; now add survivor guilt to the mix. I believe the Rangers that*

need this community the most are the ones that avoid it the most. We need to make a conscious effort to reach out to the guys we know that are not in contact, not on social media — they may build the hell in their minds to the point where it ruins the rest of their lives, and a simple 'brother, how's it going' just might be a life-saving beacon."

Thank God Cory went back. He did indeed go through tremendous difficulty when he returned to Indianapolis, and nearly became homeless for a short period. But surrounded by a supportive network, he was accepted into nursing school and became an emergency room nurse. He's a great father and was recognized at a recent mental health organization luncheon for his achievements. In March 2017, Cory was accepted into law school, and he plans to use that education to continue to help his community.

I wonder what would have happened if he'd never walked in that hanger?

I know for sure if Cory hadn't been the originator of the original "Run Ranger Run," GallantFew would not exist in its current form. His dream became a reality that has allowed us to help over a thousand veterans.

We're just beginning to scratch the surface — but we are scratching and clawing, and we are making a difference! Our opportunity and responsibility is great. Over 200,000 people leave the military each year and return to the civilian community. These men and women who put their lives on hold to serve our country and support and defend the Constitution of the United States — these people are our mission.

Karl Monger

*****

**We need *your* help to ensure they transition
from active military service
to civilian lives full of purpose and hope.**

*****

I failed to fulfill my part of Abrams' Charter. I didn't take the gift invested in me to serve in the Ranger Regiment and return it to the greater good of the Army. I've struggled with this realization for over twenty years. Ken Stauss always said, "To whom much is given much is required." He firmly believed that the privilege of being a leader in a unit such as the Ranger Regiment demanded a higher standard in excellence and sacrifice, and his love for his soldiers meant that he would do anything to ensure their success.

In sharing my story and in my work through GallantFew, I'm endeavoring to fulfill my part of the Charter now for the greater good of the veteran community. It's never too late to start. Let's dedicate ourselves to a new Charter, this one for veterans.

*****

GallantFew's Charter

Just as Rangers led the way in rebuilding the post-Vietnam Army into the world's most lethal and professional fighting force, Rangers will lead the way in rebuilding how our nation transitions veterans. We will share our lessons hard-learned with those following in our footsteps and set the example for others to follow, so men and women can transition peacefully and successfully from active military service to civilian lives of purpose and hope.

*****

> You don't have to be a Ranger to participate, just act like one. Take the initiative. Lead.

Top Ten Transition Lessons Learned

1. You're not the first veteran to transition. There is a local veteran just like you who went from where you are to success in employment and community. Find him or her and ask for their guidance. Transition will be harder than you were told it would be. Doing it on your own will make it even more difficult.
2. You come home without a strong local network. Building that network is essential for professional and emotional fitness. The perfect resume won't get you hired. That network is the best way to get a great job.
3. If you go to college, your classmates will be immature and piss you off. You'll feel like dropping out. You will need someone to help you stay focused on the big picture of graduation and employment.
4. You won't want to ask for help. You've been taught to be a leader, to "suck it up and drive on" and you will be embarrassed to admit what you perceive is weakness. Reference number 1 above.
5. You will want to hold others to the standards to which you hold yourself, and you will be frustrated and even angry when they don't care. Reference number 1 above.
6. You may have left the military not wanting to have anything to do with it in any form, but within a year you will miss it terribly and wish you were back in.

7. Isolation can feel really good—for a while—but the longer you isolate, the higher your risk for unemployment, homelessness, and suicide. Assume no one is going to seek you out to make friends; you have to take the first steps. Visiting Chamber of Commerce socials, Rotary and other civic organizations, and finding gatherings through apps like Meet Up will help you make new friends.
8. You are at risk for losing your sense of purpose. You must recreate that by intentionally seeking out opportunities to serve. Helping someone going through hardship puts your own problems into perspective.
9. Daily physical training will go out the window. Make a conscious effort to work out daily—build it into your new routine and it will stick.
10. The VA will frustrate and confound you. You won't understand how the process works and what to do to take advantage of the benefits you've earned. See number 1 above.

Bonus Lesson:

You'll be tempted to self-medicate, alcohol being the number one drug of choice. Using it to manage physical or emotional pain, to fall asleep or partying until you pass out will move you quicker towards disaster than anything else. As soon as you feel that urge, it is vital that you follow the guidance in number one above.

The Author

Karl Monger is the founder and Executive Director of GallantFew, Inc., a 501 (c) 3 not-for-profit organization formed in 2010, dedicated to helping veterans transition to civilian lives full of purpose and hope. His vision of building a nationwide network of veterans who have successfully transitioned and who guide veterans who are transitioning, has helped over a thousand veterans.

He has created The Darby Project (US Army Rangers), The Raider Project (Marine Corps special operations and infantry combat veterans), Wings Level (US Air Force), and others. He has developed or helped develop numerous innovative veteran programs to include executive producing "Prisoner of War," an award-winning veteran short film released Veterans Day, 2015. He produces and hosts the video blog, "The New American Veteran" and oversees the companion blog and podcast publication as well as the GallantFew Daily and Sua Sponte Weekly, both online newspapers.

He is certified through the Burris Institute as a Master Burris Coach (MBC), able to train and certify other Burris functional emotional fitness coaches. Karl is also owner and CEO of Transitur, LLC, a Service-Disabled Veteran-Owned Small Business specializing in personal growth coaching, veteran care and support issues, and advising corporations and agencies on veteran-related activities and practices.

As a Kansas Small Business Development Center consultant, he assisted veterans establish or expand their own small businesses. He was named U.S. Small Business Administration Region VII Veteran Small Business Advocate of the Year, 2012.

Karl began his military career as an Army ROTC scholarship cadet at Wichita State University where he earned the title Distinguished Military Graduate. He served as an infantry platoon leader, rifle company commander, and Ranger battalion staff officer, and parachuted with the 1st Ranger Battalion into Kuwait during a show of force operation in 1992.

He holds the rank of Major, retired Reserves. He is a graduate of Ranger, Jumpmaster, Airborne and Air Assault Schools, the Command and General Staff Officers Course, Combined Arms and Services Staff School, Infantry Officer Basic and Advanced Courses, the Armor Officer Basic Course, and the US Air Force Air Ground Operations School.

His awards and decorations include two Meritorious Service Medals, the Army Commendation Medal, two Army Achievement Medals, Southwest Asia Service Medal with bronze campaign star, Korea Service Medal, the Expert Infantryman Badge, Ranger Tab, Senior Parachutist and Air Assault Wings.

Since departing active duty, he has held general manager and sales manager leadership positions in several corporations and served as Executive Director of a Big Brothers Big Sisters agency. He spends his spare time keeping fit through indoor rock climbing and biking.

Karl is a past member of the US Army Ranger Association Board of Directors, past Chairman of the Board of Deacons, Eastminster Presbyterian Church in Wichita Kansas, and has mentored troubled youth and prison inmates. He is active in Metroport Veterans Association, Rotary International through his local club, Metroport Rotary, and Metroport Chamber of Commerce. He is married to the former Sara Callender and lives in Texas. He has two grown daughters and two grandsons.

GallantFew

Revolutionary Veteran Support Network

About the Organization

GallantFew, Inc. is a 501 (c) 3 nonprofit founded in 2010.

GallantFew's mission is to prevent veteran isolation by connecting new veterans with hometown veteran mentors, facilitating a peaceful, successful transition from military service to a civilian life filled with purpose and hope.

GallantFew does this by creating and supporting a nationwide network of successfully transitioned veterans who engage locally with new veterans with the same military background now going through transition, and by motivating communities throughout the nation to take responsibility for veterans returning; welcoming, connecting, and including.

GallantFew believes this will prevent veteran unemployment, homelessness, and suicide.

GallantFew, Inc. is governed by a board of directors consisting of nine individuals. Per GallantFew's bylaws, six board members must be veterans and one must be a Ranger.

Learn more at www.gallantfew.org.

Products and Services
Resources Available Through GallantFew

The Darby Project (TDP):

Serving the US Army Ranger veteran (and members of other branches of the service who work closely with Rangers) by connecting transitioning Rangers with hometown Ranger Guides. TDP also maintains a network of networks that synchronizes the efforts of veteran support organizations, to lead revolutionary transition services for the elite Army Rangers.

Learn more at www.darbyproject.org.

The Raider Project (TRP):

Serving the Marine Corps special operations (Raider) veteran and those combat Marines that bring the fight to the enemy. Key elements of the Raider Project include a WOD (Workout of the Day) app, an annual transition seminar, wilderness and climbing retreats, chartered sports fishing and a softball team.

Learn more at www.raiderproject.org.

Wings Level (WL):

Serving the US Air Force veteran, WL addresses the specific needs of these veterans. Led by a volunteer USAF special operations veteran, this program is poised for growth.

Learn more at www.wingslevel.org.

Note: Any veteran not fitting into an existing named program is managed through the greater GallantFew mentor network.

Descendants of Sparta/The Spartan Pledge:

This powerful anti-suicide message resonates with the veteran community and keeps buddies alive. Watch the video and take the pledge. www.spartanpledge.com.

Run Ranger Run:

GallantFew's annual awareness and fund-raising event, virtual teams of ten people honor Cory Smith's journey and vision and replicate his feat together. Besides providing the baseline operating funds for GallantFew, it is a powerful tool for building the network of veterans and veteran supporters globally.

Visit www.runrangerrun.com to learn more and become involved.

The New American Veteran (TNAV):

TNAV started as an audio podcast in 2010 and joined the Heroes Media Group in 2016, expanding to a video format. The purpose of the show is to drive awareness and attention to GallantFew through three main topics: highlighting veterans who have successfully transitioned and are willing to share their AAR; highlight organizations that are doing great work on the behalf of veterans; and discuss relevant veteran issues. Find links to this content at www.gallantfew.org.

Airborne Ranger in the Sky (ARITS):

This tribute site memorializes and honors every Ranger who has gone to the patrol base in the sky — whether killed in combat, training, or died of old age. The idea came from a grieving Ranger parent who contacted Karl and asked when we were going to recognize and honor his son like we did with Rangers killed in combat or in training. His Ranger had died in a hit and run accident and was not connected to the Ranger veteran community. No one knew and no one remembered.

ARITS changes that, and comrades and family can upload memories to create a lasting memorial. In April, 2017 a grandson of COL Keneally, killed in the helicopter crash mentioned several times in this book, wrote to the ARITS administrator, thanking us for giving him insight to the grandfather he never knew. Mission accomplished.

Visit www.arits.com to learn more and honor a Ranger.

DFW Fit Vet (Veteran Climbing):

Funded by a grant from the Metroport Rotary Club of Southlake, Texas, veterans climb at Summit Climbing Gym Grapevine several times a week since 2014.

Law Enforcement Education:

Karl Monger and Boone Cutler have trained over a thousand law enforcement officers at agencies across the nation on veteran issues, how to recognize a veteran, how to prevent an interaction from becoming confrontational, and how to de-escalate an interaction that has become confrontational. See www.gallantfew.org for more information.

Burris Emotional Fitness Coaching:

Every veteran joining the GallantFew network is afforded the opportunity to go through Burris Emotional Fitness coaching at no cost. We are developing a network of veterans certified as Burris Emotional Fitness coaches so that we can expand the number of veterans benefiting from this life-changing program.

What *You* Can Do

Ready to start but not sure what to do?

Veteran Community Organizing

A group of motivated veterans can accomplish anything.

Remember that US Census data shows six percent of the population are veterans, and a half-percent are under the age of 35, ergo Post-9/11. Use this formula to calculate the veteran density in your area:

Total population x .06 = total veterans
Total population x .005 = veterans under 35

Example:

A community of 20,000 times .06 equals total veterans of 1,200, and 20,000 times .005 equals 100. There are 1,100 veterans over the age of 35 who are potential Guides for the 100 who may need some guidance. That's more than a ten to one ratio.

It's up to a motivated leader to find them and organize them, that's where YOU come in.

Are there any veterans serving in leadership positions in the community? Look first to your city government. Odds are there is a veteran serving on the City Council or in a leadership position such as City Manager or Chief of Police. Set up a meeting with this person and ask if they are aware of current veteran issues. Remember the Big Three: Unemployment, Homelessness and Suicide. Ask if they are aware of any local veterans who are struggling with their transition. Encourage them to think about their friends who might have children who have served and have transitioned or are transitioning back. Talk with them about GallantFew's Guide system, and ask them to help build the network of veterans so that the community is proactive in veteran transition support. Ask them to participate in a monthly veteran breakfast. Send me an email, and I'll help you prepare for the conversation (karl@gallantfew.org).

The purpose of a monthly veteran get-together (and a breakfast is a great way to make this happen), is to provide an informal, inexpensive forum for local veterans to gather, get to know each other, and build the interconnected web of support that can engage to help any veteran living in or moving to the community.

Make it happen:

1. Identify a location.
2. Get one or two others to commit to joining you.
3. Ask the business owner/manager for their support (maybe a discount for the veterans who attend), and if you can, leave flyers.
4. Create a Facebook event and post it here: https://www.facebook.com/FirstSaturdayBreakfast/

and here: https://www.meetup.com/GallantFew/; be sure to also list it on Yelp and any other social media sites you look to for events.

5. Create a press release (GallantFew will help).

6. Contact your local newspaper, send them the press release, and ask them to list it free in their events section. You'll need to send an email to one of the editors a few weeks prior to the event.

7. Go—and no matter how many or few show up, keep going every month at the same time, and keep spreading the word. Tell those who come, to register at www.gallantfew.org as volunteers, and if they need help, have them register and ask for assistance. Let us do the verification and problem solving.

If breakfast doesn't work for the veterans in your community, make it a late afternoon event. Make it free. Don't complicate it by requiring reservations or collecting money. Don't involve alcohol. Too many veterans struggle with alcohol and use it to self-medicate; show there is an alternative.

As this started in my small community of Trophy Club, Texas, a City Council member and USAF veteran immediately understood. He stepped up and led. As our small group expanded, marvelous things began to happen. We identified a needed upgrade to our veteran memorial, and helped guide the design and development of it. By the end of 2017, there will be hundreds of veterans memorialized in stone bricks around the memorial.

We decided to hold the town's first Memorial Day event, and we learned that the small and historic Medlin Cemetery, located within our town limits, contains the final resting place of several veterans. As we spread the word and planned on the ceremony, a family member learned of our efforts and made a sizable donation to the cemetery, to improve the maintenance and upkeep.

We opened the breakfast for presentations (we do the 2nd Saturday of each month because it works well for everyone's schedule). We asked any veterans who would like to tell the story of their service to volunteer to do so. We've had several Vietnam veterans give moving talks, one with tears in his eyes because no one had ever acted like they cared about his story before—and we loved it. We are booked several months in advance with veterans eager to tell their stories, and we're planning on a large presence with a float during our town's Fourth of July parade.

We established the Metroport Veterans Association—look for us on Facebook. When you set an event, be sure to email the information to karl@karlpmonger.com.

Permissions and Credits

Reference to Steven Barber used by permission of Steven Barber.

Reference to Stefan Banach used by permission of Col. Stefan Banach, USA, Ret.

Reference to Charles Granda used by permission of Charles Granda.

Reference to Lt. Col. Glynn Hale and subsequent quotation used by permission of Col. Glynn Hale, USA, Ret.

Reference to Sgt. 1st Class Julius Kimmie used by permission of 1st Sgt. Julius Kimmie, USA, Ret.

Reference to Capt. Paul Long (posthumous) used by permission of Col Frank Long, USAF, Ret.

Reference to Jana Campbell used by permission of Mrs. Jana Campbell Truman.

Reference to Lt. Col. James M. Dubik used with permission of Lt. Gen. James M. Dubik, USA, Ret.

Reference to Col. Buck Kernan used with permission of Gen. Buck Kernan, USA, Ret.

Reference to Maj. Ken Stauss and Lt. Col. Ken Stauss used with permission of Mrs. Cathie (Stauss) Smith.

Reference to Col. David Ohle used with permission of Lt. Gen. David Ohle, USA, Ret.

Reference to Doc Donovan used with permission of Chief Warrant Officer Four Bill Donovan, USA, Ret.

Reference to Lt. Gen. Grange is used by permission of Brig. Gen. Dave Grange, Ret.

Reference to Capt. Joe Votel, Maj. Votel, Col. Votel used with permission of Gen. Joe Votel, USA.

Reference to Wesley Jurena and reprint of *Embryo Stage* in its original format are used with permission of Wesley Jurena.

Reference to Pete Parker used by permission of Peter S. Parker.

Reference to Duke Naipohn is used by permission of Duke Naipohn.

Reference to Sgt. 1st Class Michael Schlitz, USA, Ret. is used by permission of Michael Schlitz and we are grateful for his Foreword.

Quote from Josh Collins used by permission of Josh Collins.

Reference to and story of Bill Cooper is used with permission of Bill Cooper.

Reference to Big Brothers Big Sisters (http://www.bbbs.org/research/) is used with permission of Big Brothers Big Sisters of America.

Quote from Grant McGarry used by permission of Grant McGarry.

Reference to Burris Emotional Fitness and Subconscious Restructuring® by Dr. Kelly Burris is used by permission of Dr. Kelly Burris, PhD.

Reference to Beau Chatham and Warrior Life Services is used with permission of Beau Chatham.

Quote from John Risley is used with permission of John Risley.

Reference to Boone Cutler, the author of the Spartan Pledge, is used with permission of Sgt. Boone Cutler, USA, Ret.

Reference to Viktor E. Frankl *Man's Search for Meaning* is used by permission of the Fair Use Doctrine.

Reference to Killeen Chamber of Commerce study at http://killeenchamber.com/assets/uploads/docs/2nd_Qtr _2016_Report.pdf is used with permission of the Greater Killeen Chamber of Commerce and the Heart of Texas Defense Alliance.

Story of Cory Smith and "Run Ranger Run" used by permission of Cory Smith.

Back cover: GallantFew logo used with permission of GallantFew, Inc.

*****

CPSIA information can be obtained
at www.ICGtesting.com
Printed in the USA
LVOW10*1559240917
549878LV00014B/268/P